OFF
WITH
THEIR
HEADS

Also by Dick Morris

Bum Rap on America's Cities
Behind the Oval Office
The New Prince
Vote.com
Power Plays

OFF WITH THEIR HEADS

TRAITORS, CROOKS & OBSTRUCTIONISTS
IN AMERICAN POLITICS, MEDIA & BUSINESS

DICK MORRIS

ReganBooks
An Imprint of HarperCollins*Publishers*

Designed by Nancy Singer Olaguera

The Library of Congress has cataloged the hardcover edition as follows:

Morris, Dick.
    Off with their heads : traitors, crooks & obstructionists in American politics, media, & business / Dick Morris.— 1st ed.
        xvii, 343 p. ; 24 cm.
        0-06-055928-4 (acid-free paper)
        Includes bibliographical references (p. [289]–343).
            1. New York times. 2. Political corruption—United States.
3. Corporations—Corrupt practices—United States. 4. Newspapers—United States—Objectivity—Case studies. 5. War on Terrorism, 2001– .
6. France—Foreign relations—United States. 7. United States—Foreign relations—France. 8. United States—Politics and government—2001–

JK2249.M67 2003
973.931'21   2003047014

ISBN 0-06-059550-7 (pbk.)

04 05 06 07 08 BVG/RRD 10 9 8 7 6 5 4 3 2 1

To the brave men and women who have fallen, and will fall,
in defending our nation against terrorism

CONTENTS

INTRODUCTION

9/11 should have changed everything.

On 9/12, all—or almost all—Americans joined to fight terror and ban this appalling species of warfare from our planet. But as the Bush administration proceeded to act on this mandate, more and more people dragged their feet, attempting to slow the process down or hinder it from achieving its vital goals.

What happened to the unity that so blessed America after 9/11? Where did our sense of determination go? When did our focus on battling terrorism become blurred?

In the days and weeks immediately after 9/11, there was notably little dissent. As we invaded Afghanistan to take down the evil Taliban regime, virtually the entire nation supported our troops and hailed their every difficult, treacherous step through the mountains in search of those who had scarred our souls.

But then, it seemed, some of us began to lose our way. They became convinced that President Bush had "hijacked our grief," that he had "manipulated" us for his own political ends. Some began to feel that the weekly vicissitudes of the economy, and the scandals that gripped Wall Street, should be treated more urgently than stopping future acts of terror. Many demanded a miraculous settlement of the intractable Arab-Israeli conflict, even when becoming mired in yet another round of shuttle diplomacy was likely to erode the impetus for fighting the war on terror. When a settlement of the West Bank dispute proved elusive, tens of millions of Americans dissented from President Bush's decision to disarm and remove Iraq's dictator from power.

Peace marches started, as if our current troubles were somehow comparable to Vietnam. In some quarters, George W. Bush was more reviled than Saddam Hussein. We became a nation divided.

Demonstrators, who might have better spent their time outside the

Iraqi embassy demanding disarmament, flocked instead to march in front of the U.S. embassy, begging for inaction.

This book is about those political, journalistic, and cultural leaders who have mounted a campaign to oppose and impede the war on terror that seemed so vital in the rare moment of clarity we shared after 9/11. As a New Yorker, I find 9/11 is still very fresh in my mind. I feel the absence of the missing towers of the World Trade Center like I would molars extracted from the back of my own mouth. In the moments after the terrorist attack, almost all of us silently vowed to do what we could to stop this slaughter from happening again. As some have lost sight of that commitment—or forsworn it altogether—our momentum in battling terror has slowed.

As the days have unfolded since 9/11, the forces that cast obstacles in our path have proliferated; those who are working to carry through the resolutions we made on that horrific day are at risk of becoming overwhelmed, outnumbered, and outmaneuvered. This book represents an effort to expose the distortions, obfuscations, and sleights of hand with which opponents of this war on terror seek to mislead us all.

If America is truly united, we can slay any dragon; witness the way our nation was galvanized to defeat the threat of fascism during World War II. But we are a nation that insists on embodying its values even as we fight to defend and promote them. We have a strong national conscience, and sometimes, as Hamlet says, "conscience does make cowards of us all." Our conscience must drive us to defend our values at the time of maximum danger, not strip us of the will to action.

Even as we confront our enemies abroad, we must also be vigilant about those who sap our will from within, undermining our resolve and our unity. They do so through no treasonous intent, nor does this book suggest that their voices should be stilled. But they must be answered, with all the strength and commitment our war for self-preservation demands. For in exercising their constitutional rights of free expression, they might make cowards of us all, if their arguments go without rebuttal and their criticisms without a parry.

And, finally, as we look within our borders, we must remain equally alert to threats that may seem secondary to the war on terror—but that nonetheless, in their various ways, weaken the welfare, the economy, and the very democracy we are fighting to preserve.

Sounding a warning call, on all these fronts: that's what this book is all about.

To put it more personally: What I take on in these pages are people, forces, and institutions that make me mad. A lifetime in politics may have made me cynical, but I'm not numb. When I see distortions and deceit dominating our political dialogue, it drives me crazy. This book is my personal *cri du coeur* about deception in politics, journalism, and business—especially when it stops us from following through on the work 9/11 has left for us all to do.

This book takes on some pretty sacred cows, but it's about time they became fair game.

In Part I, I take on those who are overtly hindering our efforts to wage war against terror. Not because they are unpatriotic. Not because they wish us ill. But because they are wrong.

First I go to the leader of the opposition, the drummer whose cadence governs the step of the march against President Bush's battle to fight terrorism—the *New York Times.*

I still pick up my copy every day, as I have since I was eight years old (although back then I turned to the sports pages first to see how the Yankees were doing). But it is no longer my father's *New York Times*—impartial, fair, understated, and reliable. It's a new kind of *New York Times*, remade by managing editor Howell Raines, who took over just days before 9/11, and I don't like it one bit. It has turned from a tower of rectitude into the Leaning Tower of Pisa—leaning left.

Each month, the *Times* trots out a new line in its efforts to counter Bush: the civil liberties of the captured al Qaeda, the rights of Muslim immigrants questioned by the FBI, the priority it says we should give to the economy. The left picks up on its signals like football receivers watching their quarterback. The *Times* sounds the theme, and a thousand liberal groups perform their own variations on it. Its editorial biases, slanted news coverage, weighted polling, and persistent repetition of its stories go out like a daily talking-points message to Democrats and the left. But the signals come not from a political party but from a newspaper that is supposedly impartial.

When other newspapers distort or propagandize, some of us may go astray for a few moments. But when our national newspaper of record does it, we are all thrown off course. It's as if the compass no longer points north but, instead, points left. The television networks and other establishment media are just as bad as the *Times*. During the war with Iraq, we saw just how biased they really are—constantly hyping their point of view at the expense of objectivity. The truth doesn't seem to matter. Only their collective ideology and predispositions are important. They shine the news through a prism, so it comes out in precisely the colors they want.

Bill Clinton and his record in the war on terror is my next target.

Clinton has had a tough ex-presidency. The pardons and the payoffs, the gifts and the guilty pleas: that was one thing. But now the much more dangerous legacy he left us is coming clear: his disastrous record in "fighting" terrorism. Anyone who saw 9/11 realized that Bill Clinton had failed to prepare his nation for an imminent terrorist threat.

But that horrible day turned out to be just the beginning. Soon we learned more: how his "agreement" with North Korea to curb its desire to acquire nuclear weapons was flawed and unenforceable. How Iraq had taken advantage of the expulsion of U.N. inspectors, and the elimination of the ban on its oil sales, to finance and build armaments, including weapons of mass destruction. These concessions, granted by Bill Clinton, rearmed Iraq and made it again a threat to the world.

Clinton left these time bombs—al Qaeda, Iraq, and North Korea—ticking under the White House, knowing full well that they could explode on his successor's watch. His motto might well have been, *Après moi, le deluge:* After me, the disaster.

Don't misunderstand: This isn't about digging up ancient history. It's about investigating the mistakes Clinton made—about uncovering where the time bombs are buried, when they're likely to explode, why they were put there, and how to dig them up.

In my earlier book *Behind the Oval Office,* published in 1997, four years before 9/11, I recounted how I told Clinton in 1996 that I felt that his place in history depended on "breaking the back of international terror by military and economic means." But he would have none of it. Deny driver's licenses to illegal aliens, and give highway cops the technology and data to pick up those here illegally? No way. That might "lead to racial profiling." Issue a list of charities that raised money for terrorist groups? Nope. Attorney General Janet Reno thought it would be too much like McCarthyism.

Enforce sanctions against Iran so foreign companies don't subsidize its oil production? So Iran can't fund terrorism or acquire nuclear weapons? Negative. Deputy National Security Advisor Sandy Berger thought it would antagonize Western Europe. Require photo identification to board planes? Federalize airport security workers? X-ray all checked luggage? Nyet. Vice President Al Gore declined to put them in his recommendations for aircraft safety and security.

On welfare reform, balancing the budget, fighting crime, and a dozen other topics, Bill Clinton was active, astute, aware, and alive. On terrorism, he was AWOL. And now we are paying for it.

But Bill Clinton isn't the only one who has failed us in the war against terror. Some of our most admired writers, journalists, actors, actresses, social commentators, humorists, directors, pundits, and others have become the **Hollywood apologists,** jumping on the bandwagon to blame the American people and their government for 9/11, and they are opposing our efforts to disarm Iraq and protect the safety of the world. Their rationalizations may sound principled—but even a cursory examination reveals that they're based on pure ignorance.

These icons of stage and screen, song and dance, are no longer content with making us tap our feet. They want to change our minds. But they bring to their advocacy their old habits—they follow their scripts. They are not intellectuals. They are actors, actresses, singers, and stars who are impersonating deep thinkers. The same skills they use to persuade us that their stage characters are really in love, or truly locked in mortal combat, they now employ to try to convince us our country is going in the wrong direction. Their skills are formidable. But let's not forget the reality: These are human parrots, mouthing lines fed to them by the fashionable, social, trendy elite. Their information is as shallow as their conclusions are vacuous.

The danger is that some of us might be deluded into following these stars as we navigate our way through the seas of international terrorism. Accustomed to following them in their crafts, we might mistake the celebrity for the cerebral—and be led down the garden path to appeasement or defeat. They claim the United States has blood on its hands, that it deserved to be a target of terror. They claim we are trading blood for oil. Some even say that we are becoming terrorists ourselves in the name of fighting terror.

So here the apologists are: Gore Vidal, Woody Harrelson, Dustin Hoffman, Susan Sarandon, Tim Robbins, Barbra Streisand, Andy Rooney, Bonnie Raitt, Sheryl Crow, George Clooney, Danny Glover, Senator Russ Feingold, Sean Penn, Susan Sontag, Norman Mailer, Rev. Jerry Falwell, David Clennon, Bill Maher, Noam Chomsky, Adlai E. Stevenson III, Ozzie Davis, Ed Asner, Gerda Lerner, Alice Walker, Barbara Kingsolver, Grace Paley, Eve Ensler, Tony Kushner, Laurie Anderson, and . . . President Bill Clinton. Their arguments are based on emotional, illogical, unfounded clichés, which contradict the facts at every turn.

Don't buy any of it—not for a minute.

The perfidy of our oldest ally—**France**—draws my special scorn. I've always loved France and visit it several times a year. I even worked hard to learn to speak French. My wife's grandfather was wounded twice defending the French people: a twenty-year-old Irish immigrant at the time, he joined the American army and fought at the Battle of Château-Thierry in World War I. His citation hangs on our wall. A visit to the Normandy D-day graveyard never fails to move me to tears.

But, now, when we are under attack, when it is *our* lives and *our* nation that are in danger, where are the French? Doing their best to trip us up, foil our efforts, tip off our opponents, and aid those who are our enemies. Indeed, it is they who helped create the monster Saddam Hussein, leading the fight to allow Saddam to sell as much oil as he liked and do as he pleased with the money.

France is no longer an ally of the United States. When our backs were turned and we were at our most vulnerable, they proved a fickle friend and a selfish sister. Regardless of the sacrifices we made to save them in two hot wars and one cold one, when we needed their help, they said *"Mais non."*

In **Part II**, I go after those who have attacked us from within.

Let's start with those who led the **attack on our economy**. The recession that gripped America after 9/11 has weakened our ability and undermined our will. We were all so worried that we stopped flying, buying, and investing. The result? The funds we need to fight terrorism are no longer flowing freely through the adrenaline system of our economy.

But Osama bin Laden did not cause our recession single-handedly. He had help. The corporate executives at Enron, Arthur Andersen, Global Crossing, WorldCom, and a host of other companies chose that moment to pull off the greatest robbery in history, which further tested our confidence as it set the stock market plunging.

We all know the story of how Enron misrepresented its revenues and profits and how the accounting firm of Arthur Andersen taught Enron how to lie and get away with it. But that's not the full story. Why didn't the Securities and Exchange Commission stop the shenanigans? And why couldn't the investors intervene either to prevent the larceny or, at least, to get their money back?

The answer lies not just with these corrupt firms but with two leading politicians—Senators Chris Dodd (D-Conn.) and Phil Gramm (R-Tex.), who cut a political deal that paved the way for the Enron debacle and the other scandals. Both parties were in on it. And President Clinton followed along, pretending he was opposed.

How do I know? I was there.

The reform legislation of 2002 will do nothing to help those who have lost their life savings get it back. All the laws that protect the corrupt and punish the misled are still on the books, and they will make it impossible for individual investors to achieve justice. Until these laws are repealed, Washington may reform Wall Street for future investors, but those who have imperiled their retirement by investing in the 1990s will have no hope of redress.

Then there are those party bosses who led the **attack on our democracy.**

Not all the threats to our democratic way of life fly airplanes into buildings or create weapons of mass destruction. These terrorists may seek to destroy the Capitol in Washington from the outside, but the incumbent congressmen who work there have shown that they're just as intent on destroying our democracy from within.

Unbeknownst to the American people, before the 2002 election, incumbent congressmen of both political parties got together to reapportion their districts so craftily that they, in effect, denied us the right to choose anyone but them for Congress. They took the power of the vote right out of our hands, gerrymandering themselves into a life tenure in office. And they did it while we weren't looking—while we were bracing ourselves for the battle against terror that will define our generation.

As a result, 96 percent of incumbents got reelected—and some who were beaten lost to other incumbents!

The way they've left things, 415 members of the House of Representatives are all but chosen before you get into the voting booth. Only one American in twenty gets to cast a ballot for the handful of seats that are really contested. They call that democracy.

And politicians wonder why turnout is dropping.

My final two targets are those who exploit the very young and the very old. They repeat the pattern of 9/11, punishing the innocent and rewarding the guilty. Their inhumanity saps our national virtue and will.

No terrorist ever killed as many innocent Americans as those who **attack our children**—cigarette company executives—do each year. Bin Laden is small-time next to the owners, employees, and directors of the cigarette companies. But lately they have had a partner in killing American adults and addicting our children: state governors who are squandering the tobacco settlement money on their big-spending schemes, sacrificing public health in the process.

It didn't have to happen this way. The amazing tobacco settlement between the state attorney generals and the cigarette companies in the 1990s promised to usher in a new era in public health, promising to pump $9 billion a year from tobacco companies into antiteen smoking programs. The results were sensational, better than anyone had hoped. Lung cancer deaths dropped 14 percent in California, where the antismoking effort was most intense. Teen smoking dropped by 20 percent. Death was losing its grip on hundreds of thousands of us.

Then came the politicians—like California's governor Gray Davis—to spoil it all. Pressed for cash as his state deficit mounted, Governor Davis diverted the tobacco windfall—money that could have been used to stop kids from smoking—and squandered it on big-spending programs instead. Anxious to avoid higher taxes, he degraded the health of his state instead. Now, tens of thousands more kids will smoke, and adults will die in pain, all because Governor Davis needed to balance his budget.

At the other end of the aging spectrum are the vicious entrepreneurs who **attack our elderly**. There is no terror worse than that felt by elderly people locked in nursing homes, supervised by abusive employees and absent owners. America's nursing home owners are offering worse care for more money than ever before. They keep 3.5 million of our parents in hellholes that offer bad service and little care, and invite death by starvation, dehydration, bedsores—and even, recently, arson.

The federal government estimates that 92 percent of nursing homes are understaffed. The record is replete with instances of massive physical abuse, rape, torture, verbal tongue lashings, beatings, and even murder inside our nursing homes. Congressional studies indicate that actual abuse of residents—not just neglect, but *abuse*—takes place in one-third of America's nursing homes.

Moreover, the two-thirds of all nursing homes that are owned for profit are raking in outrageous profits, at the expense of care for those who live in them.

To ensure that they get away with this mayhem, the nursing home owners have cooked up a deal with the politicians to make litigation against homes almost impossible in many states. So the abuses flourish, and neither the courts nor Bush do anything about it.

And, so, to these men, women, companies, and countries who abuse the public trust in business, the media, and politics, I say, with apologies to Lewis Carroll's Queen of Hearts:

OFF WITH THEIR HEADS!

PART I

THE OBSTRUCTIONISTS

As the United States goes to battle against terrorism, there are many among our countrymen and our traditional allies whose efforts are threatening to drain our spirit and determination, attacking our actions as we wage this war.

Some have attacked their motives, or denied their right to dissent. Not me. My problem with them has nothing to do with their right to dissent, which is beyond question. The real problem is simpler, and far more dangerous: their premises are misjudged, their comments based on ignorance, their arguments simply wrong.

Sometimes, as with the *New York Times,* their obstructionism comes veiled behind a facade of objectivity. Bill Clinton's obstructionism was not an overt act, but a failure to act when the power, the opportunity, and the necessity for action pressed upon him. The apologists who denounce America as we seek to rid the world of the scourge of terrorism are overt and active in their opposition. They just don't know what they're talking about. France, our traditional ally, has turned into our overt adversary, blocking, inhibiting, and countering our efforts to stop Saddam Hussein from terrorizing the world.

But all four—the *Times,* Bill Clinton, the apologists, and the French—have this in common: They have put our nation at jeopardy.

ONE

THE NEW *NEW YORK TIMES:*
ALL THE NEWS THAT FITS, THEY PRINT

There is a new *New York Times*. Howell Raines's *New York Times*. No longer content to report the news, he admits to "flooding the zone"—and floods it with stories that carry forward his personal crusades and the paper's editorial views.

And the *Times* doesn't stop at slanting the news; it also weights its polls. The surveys the newspaper takes regularly are biased to give more strength to Democratic and liberal opinions and less to those of the rest of us.

The newspaper has become like a political consulting firm for the Democratic Party. Under Raines, it is squandering the unparalleled credibility it has amassed over the past century in order to articulate and advance its own political and ideological agenda.

For decades, the *Times* was the one newspaper so respected for its integrity and so widely read that it had influence well beyond its circulation. Now it has stooped to the role of partisan cheerleader, sending messages of dissent, and fanning the flames of disagreement on the left. Each month brings a new left party line from the paper, setting the tone for the government's loyal opposition.

Reading the *New York Times* these days is like listening to Radio Moscow. Not that it's communist, of course, but it has become almost as biased as the former Soviet news organ that religiously spewed the party line. Just as Russians did under Soviet rule, you now need to read "between the lines" to distinguish what's really happening from what is just *New York Times* propaganda.

I have read the *New York Times* for forty-four years. When I was growing up, my parents read it every morning and the *New York Post* every afternoon. I still read them both every day. The *Times* is a New York institution to me, as much a symbol of my hometown as the Yankees, the subway, Central Park, and, yes, the World Trade Center. I think many Americans must share my feelings today: To see it fall into the hands of propagandists, after so many years of dignity and balance, is like watching your father get drunk.

Like every newspaper, the *Times* rightfully uses its editorial and op-ed pages to articulate its ideas and opinions. But, since the ascension of Howell Raines to the post of managing editor, the newspaper has gone much, much further to push its political perspective. As journalist Ken Auletta pointed out in a masterful profile in *The New Yorker*, Raines is overt about his desire that "the masthead" (the managing editor, his deputy, and the assistant managing editors) "be more engaged in shaping stories and coordinating news coverage."

Acting like the chief campaign strategist for the left, the *Times* generally conducts six to eight public opinion polls each year. But lately the *Times* seems to me to be deliberately misinterpreting and weighting its data to suggest that its liberal ideas have a popularity they don't actually enjoy. The polling seems to have one major purpose—to help the Democratic Party set its agenda, encouraging it to embrace the *Times*'s own liberalism on a host of issues. Then, from editorials to op-ed articles and a blizzard of front-page stories, the newspaper relentlessly expounds its views, doing its best to create a national firestorm on the issues it chooses to push.

Jack Shafer, the media critic for the on-line magazine *Slate*, described the new policy to *Newsweek* on December 9, 2002: "The *Times* has assumed the journalistic role as the party of opposition" to the current Bush administration. According to *Newsweek*, "many people around the country are noticing a change in the way the Old Gray Lady [the *Times*'s pet name] covers any number of issues. . . ." The magazine pointed out how Raines believes in "flooding the zone—using all the paper's formidable resources to pound away on a story."

Other newspapers often try to do the same thing. What is unique about the *Times*'s approach is the sharp departure it represents from the paper's past. Long priding itself on objectivity, political neutrality, and even reserve in reporting news, the *Times* is renowned as our nation's primary voice of objective authority. As such, it occupies a unique place in our national iconography. But Alex Jones, author of *The Trust*, a book about the *Times*, describes the *Times*'s latter-day style of news coverage as "certainly a shift from the *New York Times* as the 'paper of record.'"

And yet millions of us still rely on the *New York Times*. It is still the most comprehensive source of news and information about what is going on in the world. It is precisely because it is so important that the bias that increasingly dominates its coverage of news is so disturbing. It's a little like finding propaganda in the *World Almanac*—the place you want to go to get the facts and only the facts. If we cannot depend on the *Times* to tell us fairly, accurately, and dispassionately what is going on, where are we supposed to turn? Will news reading become a task in which we must read four or five partisan sources and average them to get the truth? Is it really worth subverting an institution like the *New York Times* just to score political points?

Every day's front page is such a mix of hype, hyperbole, and, often, hypocrisy that it takes an expert to sort it out.

While most nations have their national newspapers, American newspapers, with the exception of *USA Today* and the *Wall Street Journal*, are all local in orientation. The *New York Times*, however, leads a double life—as the most widely read newspaper in the nation's largest city *and* the most authoritative voice on national news. Seen as a national tower of rectitude, the *Times* has always enjoyed universal respect for its even-handed impartiality.

So it's not surprising that the impact of this *Times* propaganda offensive was far more widespread than its daily circulation of 1.1 million would suggest. Not only do most opinion leaders in America read the newspaper itself, but the *New York Times* News Service—the paper's equivalent of the Associated Press—sends stories to scores of other daily papers around the country. In addition, its stories are reprinted in the *International Herald Tribune* and disseminated in every major city in the world, and, of course, are available on the Internet.

Beyond the nominal reach of the paper and its wire service, however, the themes set in the *New York Times* are crucial in shaping trends in journalism throughout almost every paper in the nation. During my time in the Clinton White House, I tracked carefully the themes that were covered on the front pages of twenty newspapers in swing states throughout the nation. Each week my staff detailed the topics covered in such diverse dailies as the *Cleveland Plain Dealer, Chicago Tribune, St. Louis Post-Dispatch, San Francisco Examiner, Miami Herald,* and other pivotal papers in key states. In addition, we evaluated the number of minutes each of the three networks devoted to each news topic.

That ongoing survey revealed just how closely the themes covered in print and on TV tracked those first articulated on the front page of the *New York Times.* When the *Times* spoke, ripples seemed to flow out from the initial news splash it made, touching scores of other, more local, news organs.

I once asked George Stephanopoulos why he thought so many other venues tracked the *Times* so closely. "The *New York Times* still rules," he replied.

No longer content simply to report the news, the *Times* now seems to want to make and shape it, focusing attention on issues and topics that advance its liberal views. In the period since 9/11, the *Times* has been particularly active, grappling with the conundrum that had liberals flummoxed all year—how to respond to the growing national conservatism in the face of the al Qaeda attack.

Facing a Republican president with record-high approval ratings, determined to bring terrorism to heel, the *Times* puts its polls, its editorials, and its front page to work, marshaling one strategy after another to regain the momentum the left lost after September 11.

Using polling to feel its way—just as a candidate for president might—the newspaper seems to consciously choose a particular political strategy and then uses all its resources to beat the drum persistently—manipulating its story placement, choice of photos, headlines, article topics and language, editorials, and op-eds to push its point of view. If the specific issue chosen by the *Times* isn't at the top of the public's mind, the newspaper resolves to put it there, by running daily front-page articles with bold headlines, slanted "push polling," and urgent editorials. And when that doesn't work, it simply gives up and chooses new targets.

Mickey Kaus of *Slate* has accused the *Times* of moving away from its classical mandate to "follow the news" to a "mandated-from-above-throw-the-whole-staff-into-the-story news *campaign.*"

Kaus attributes much of this development to the rise of Howell Raines to the position of managing editor at the *Times*—on September

5, 2001, just six days before disaster struck. Kaus asserts that Raines has injected his "slightly intemperate, self-righteous populism" into both the Editorial Page and the paper's news coverage.

Raines, fifty-nine, is a self-styled southern populist, who, according to *Newsweek*, "cut his teeth at a time when the Southern papers were still charging the barricades of segregation." His early career was marked by an assault on the racist values that dominated his hometown of Birmingham, Alabama. (Bill Clinton, chafing under the *Times*'s editorials attacking him, told me that Raines has it "in for me because I'm a Southerner who didn't have to leave to make good.")

A confident, unapologetic liberal, Raines sees journalism as an opportunity to crusade by capturing a story and pushing it into the public consciousness—again and again and again. Strutting around Manhattan, ostentatiously clad, as Auletta describes it, in the "white panama hat of a plantation owner," he insists on putting his own imprint not just on his newspaper but on the minds of its readers.

As Raines told Auletta, "Target selection is key. And then you have to concentrate your resources at the point of attack. [John] Geddes [the *Times*'s deputy managing editor] has another term for my style, which is 'flood the zone.' I've been in journalistic contests where I was up against real formidable opposition. . . . If I'm in a gunfight, I don't want to die with any bullets in my pistol. I want to shoot every one."

At the root of the *Times*'s "flood the zone" strategy is an adroit use of opinion polling to develop a national strategy and to convince politicans—particularly those on the left—to embrace it as their own. Borrowing heavily from the techniques of Clinton's permanent campaigns of the 1990s, the newspaper used survey research to identify the hot-button news issues and then developed them into stories, features, and editorials.

But a close analysis of the *New York Times*'s polling reveals that the data itself was far from unbiased; indeed, it was often weighted, twisted, and slanted to advance the newspaper's liberal point of view. Raines's *Times* used techniques reminiscent of "push polling"—where consultants stack the questions in surveys to give the impression of more public support than they, in fact, can muster—to get the results it wanted.

OFF WITH HIS HEAD!

Push Polling at the *Times*

As Mark Twain once wrote, there are three kinds of lies: "Lies, damn lies, and statistics." It's a sentiment that cannot have gone unnoticed at Howell Raines's *New York Times*.

Going beyond simple news coverage, the *Times* joins with the equally liberal CBS television network's news department to commission periodic surveys of public opinion. These polls, conducted by telephone, are typically reported in front-page stories, and they tend to shape the political debate for weeks afterward.

Most senators and congressmen take polls only during election campaigns. Even then, their surveys tend to focus on their personal approval ratings rather than on broad issues of public policy. And, of course, the surveys are conducted only in their particular states and districts; they don't involve national samples.

While today's White House probably polls frequently—I polled every week for Clinton—it generally doesn't share its findings broadly within its own party and obviously never gives data to the other side. The Democratic and Republican National Committees poll from time to time, but the average member of Congress has little access to national polling, except for what he or she reads in the newspapers. As the most prestigious of all news organs, when the *New York Times* announces a poll, politicians everywhere sit up and listen.

Knowing its power, the *Times* has lately begun using polling techniques pioneered by partisan political survey research firms, testing themes for liberal candidates and probing for weak spots in conservative positions and the Bush administration's image.

There's only one problem: The *Times*'s polls are slanted!

A close evaluation of the newspaper's polls between 9/11 and Election Day 2002 reveals that the *Times* weights its data artificially, tilting its numbers to the left.

Here's how it works: The newspaper's pollsters interview Democrats, Republicans, and Independents to conduct the survey. Then they weight *up* the Democratic responses and weight *down* the Republicans', pushing the numbers to the left by between one and five points in each survey.

Weighting isn't always wrong. It's often a valid way to correct for errors in the sampling. Surveys are conducted by telephoning a random sample of voters. Generally, those conducting the survey don't know if the people they are calling are Democrats, Republicans, or Independents, young or old, rich or poor, black or white. When the results of the survey come back, the pollster often finds that he or she has too many of one group and not enough of another. So the practice of weighting was developed to rebalance raw data to reflect more accurately proportions of gender, race, religion, age, and place of residence.

For example, it's common for more women to answer the telephone than men. So a survey might include responses from more women and fewer men than it should. In that case, it would be perfectly legitimate to weight the data to correct for this flaw in the sample.

Suppose a survey of a thousand respondents turns out to include 550 women and 450 men instead of the 500/500 ratio it should have. A legitimate pollster would adjust for this sampling error by giving each male interview 10 percent more weight and each female 10 percent less; this would create the same result as if he or she had actually interviewed 500 from each gender.

The *New York Times* weights its data as any other polling operation does. But between January 1, 2002, and the November election of that year, each one of its polls weighted *up* the number of Democrats and weighted *down* the number of Republicans—every single time!

If the *Times* were using weighting to adjust for sampling error, surely the weightings would sometimes increase the number of Democrats and would sometimes decrease it. But what the *Times* has done—increasing the ratio of Democrats to Republicans each time—isn't weighting the sample. It's slanting it.

The following table demonstrates the weightings the *Times* used in its polling. At the left is the date of the *Times*'s survey. The second column breaks down the number of actual interviews conducted among Democrats, Republicans, and Independents. The third column shows the weighted distribution among the three groups; the final column, on the right, shows the difference plus or minus as a result of the weighting.

What's the bottom line? In every single survey, the number of Democrats interviewed is weighted *up*—by between 0.4 percent and 2.4 percent—while the number of Republicans interviewed is weighted *down*, by between 0.8 percent and 3.1 percent!

The *New York Times* Weights Its Polls

Date	Actual Interviews	Weighted Interviews	Difference
Dec. 7–10, 2001			
Dem	329 (31.3%)	333 (31.7%)	+0.4%
Rep	339 (32.2%)	319 (30.4%)	-1.8%
Ind	384 (36.5%)	399 (38.0%)	+1.5%
Total	1052 (100%)	1051 (100%)	
Jan. 21–24, 2002			
Dem	328 (31.7%)	346 (33.5%)	+1.8%
Rep	307 (29.7%)	294 (28.4%)	-1.3%
Ind	399 (38.6%)	394 (38.1%)	-0.5%
Total	1034 (100%)	1034 (100%)	
July 13–16, 2002			
Dem	318 (31.8%)	342 (34.2%)	+2.4%
Rep	310 (31.0%)	284 (28.4%)	-2.6%
Ind	372 (37.2%)	373 (37.3%)	+0.1%
Total	1000 (100%)	999 (100%)	
Sept. 2–5, 2002			
Dem	317 (33.8%)	330 (35.2%)	+1.4%
Rep	280 (29.9%)	273 (29.1%)	-0.8%
Ind	340 (36.3%)	335 (35.7%)	-0.6
Total	937 (100%)	938 (100%)	
Oct. 3–5, 2002			
Dem	240 (35.9%)	252 (37.7%)	+1.8%
Rep	201 (30.1%)	180 (26.9%)	-3.2%
Ind	227 (34.0%)	236 (35.3%)	+1.3%
Total	668 (100%)	668 (100%)	

See what's happening here? In the poll of January 21–24, the actual sample turned up 328 Democrats (or 31.7 percent of the sample) and 307 Republicans (or 29.7 percent of the sample). But in reporting the results, the *Times* counted each Democratic interview as more than one and each Republican as less than one, as if there had actually been 346 Democrats (33.5 percent) and 294 Republicans (28.4 percent). The result was to make the poll 3.1 percent more Democratic than it would have been before the weighting.

Obviously, this weighting had a significant impact on the public message of each new *Times* poll—from understating Bush's approval ratings to overstating doubts about a U.S. invasion of Iraq. The weighting, since it is applied to every one of the questions on the *Times* surveys, skews all the answers to the left.

But wait, you may be thinking: *What if the* Times *surveys just happened to get too many Republicans and too few Democrats each and every time they polled?* Wouldn't it be justifiable to weight the data to increase the Democrats each time? Would that really be bias, or just a consistent correction of data that happened to have the same flaw each time?

No. Look back at the table on the previous page. Notice how each poll has a different number of Democrats and Republicans after it is weighted.

In December, the *Times* weighted its data to get 31.7 percent Democrats and 30.4 percent Republicans. The next month, it ended up with 33.5 percent Democrats and 28.4 percent Republicans. In July of 2002, its weighted survey turned up 34.2 percent Democrats and 28.4 percent Republicans. And so on. Each month's survey produced a different final number of members of each party.

Did the *Times* presuppose that the number of Democrats and Republicans varied so continuously? No, it couldn't have. There's no data to support that conclusion. What happened was that each month the newspaper weighted its data to make its sample more Democratic and less Republican *because it wanted to*, not because of any preconceived target.

When the pollsters and the editors at the *Times* read this, they will, undoubtedly, come back and say that they don't weight for party—they just weight for demographics. They will attribute the change in the partisan mix of their sample to the unintended by-product of a weighting to get proper demographics.

For example, they might say that their telephone polls get too few blacks and Hispanics, forcing them to weight up the number of minorities to get a valid national sample. Of course this makes the entire sample a bit more Democratic, they will say, but that isn't their intent, it's just a by-product.

This argument would hold water except for two things. First, the *Times*'s published polls give information only on weighting by party; there is no indication that they weight for race or other demographics. Second, their weighting has *always* made the sample more Democratic in every poll since 9/11. If even once in a while these unintended consequences made the poll sample more Republican, the demographics argument might make sense. But the polls' own wording is their own smoking gun.

Pollster John Zogby observes that when he weights his national samples they frequently become more Republican. "All of us in the polling profession apply weights to our samples to better represent demographic sub-groups. I also apply weights for party identification because I have found that that those groups generally more inclined to vote Democrat are usually more willing to answer the phone than some Republican sub-groups. This has worked well in national elections. I have always been curious why some pollsters purposely over-represent Democrats. I think it is a mistake."

The fact that the *Times* always managed to make its sample more Democratic indicates more than a random by-product of demographic weighting. In my opinion, it suggests a deliberate effort to bias the results.

The effect of this artificial weighting is not only to confound the *Times*'s enemies but also to fool its friends. By overstating the liberalism of the electorate on issues such as Iraq and immigration, the *Times* has advanced its own agenda with its weighted samples—but in the process it has misled those most sympathetic to its views. Throughout 2002, Democratic politicians relied on *New York Times* polling data to move to the left on key issues, only to find out through Bush's victory in the midterm elections that they had been following a false prophesy.

The *Times* not only biases its polls by weighting the sample, however. It also appeared to use the text and sequence of its questions to lead respondents toward the answers it wants to hear.

One of the most fundamental concerns of pollsters is not to "pollute" the sample of respondents so that one question biases and predetermines the answer to the next one. Pollsters work hard to prevent the respondents from knowing which side of the issue is sponsoring a poll, because they know that most respondents treat the interviewer as a figure of some authority and are influenced by what they perceive to be his or her opinions.

But those seeking to bias a survey can exploit both of these phenomena, to slant the outcome in their direction.

A review of the *New York Times*'s coverage of the war on terror from 9/11 through Election Day 2002 reveals a consistent pattern throughout the period. Raines and his staff would take the surveys and then frame their report of the results to confirm the paper's editorial position. Even when the results diverged significantly from the *Times*'s position, the journalists reporting on each survey—and the headline writers in particular—would search through the data, sifting it finely, to find questions, even if taken out of context, that appeared to validate the paper's editorial views.

Then would follow a blizzard of news stories, editorials, op-eds, and news analysis pieces stoking further the liberal agenda the newspaper had laid out in its polling. By hammering away at certain familiar liberal themes every day and using biased polling to report the public's agreement, the newspaper became an influential propaganda organ that put its own liberal stamp on the national political agenda.

Like Howell Raines's own one-man band, the new *New York Times* had found a way to sing with one voice. It took the polls, published the results, wrote the editorials, reported the news to suit its views—and then polled again to document how the public agreed with its chosen position.

After 9/11—Laying Out the Liberal Agenda

As New York City dug its way out of the rubble of 9/11 and its battered residents recovered from their shock, the *New York Times* took the lead in reporting on the tragedy; deservedly, it won a Pulitzer Prize for its reporting on the attack. To its credit, each day the paper printed one or two full pages crammed, top to bottom, with the photos and encapsulated eulogies of the dead. The effect of this roll call of the innocent was overpowering; it represented journalism at its best. Sometimes it seemed as if the entire newspaper was an obituary page. One could almost hear it crying.

Meanwhile, however, Raines's *Times* was working to set out a liberal agenda to respond to the attack. In its news stories, and particularly in its polling, the newspaper sounded, tested, and refined the themes that were to constitute the left's response to the devastation of the terrorist attacks.

As the *Times* scrambled to set the liberal agenda in the aftermath of 9/11, it developed themes and approaches that would become more and more pronounced as the next eighteen months unfolded.

A diligent survey of its news coverage reveals seven distinct phases in the *Times*'s increasingly frantic efforts to break the conservative momentum after 9/11. The pattern was always the same, following the Raines formula: flood the front page with stories and photos, push the party line on the editorial and op-ed pages, and gin up the impression of public support through weighted polling to show how popular the issue du jour was politically.

Themes of the *Times*'s Post-9/11 Coverage

Date	Theme
Sept–Oct, '01	Huge Losses Likely in Afghan War
Nov–Dec, '01	War on Terror Is Threat to Civil Liberties
Jan–Feb, '02	The Enron/Wall Street Scandal Is Now the Top Issue
Mar–May, '02	Resolving the Arab-Israeli Conflict Is Key to the War on Terror
June–July, '02	Economic Slump More Important Than Foreign Issues
Aug–Sept, '02	Rein In U.S. Action on Iraq
Oct, '02	Economy Is Key Issue, Not Terror

September–October 2001

The Times *Focuses on Dangers of War*

Shortly after 9/11, the *New York Times* began to stress how risky the war in Afghanistan in particular, and against terror in general, was likely to be. Emphasizing the probable high cost, in both lives and money, of an American military strike against the Taliban, the newspaper ran twenty-five front-page stories in the fifty-one days between 9/11 and the end of October 2001 stressing the difficulties and dangers of a war in Afghanistan.

In the newspaper's initial coverage of the attack, it reported, in a front-page headline: "Washington and Nation Plunge into Fight with Enemy Hard to Identify and Punish." Each day in September, the *Times* reported new difficulties facing our forces in the expected attack on the Taliban.

- "Scarcity of Afghanistan Targets Prompts U.S. to Change Strategy"
- "Peacetime Recruits Getting Ready for War's Perils"
- "The Tough Afghan Terrain"
- "A Guerrilla War (Looms)"

In October, as military action approached, the headlines assumed an even more pessimistic tone.

- "Ground Raids Seen as Long and Risky, British Admiral Predicts Ground Fighting and Months of Strikes"
- "Insertion of Ground Troops Demonstrates Willingness to Risk American Casualties"
- "Allies Preparing for a Long Fight as Taliban Dig In; Optimism of Early October Fades . . ."

The newspaper painted a grim picture of the frustrating efforts the Red Army had encountered years before, trying to defeat the Taliban fighters our troops would be facing. It deprecated the capacities of our Northern Alliance allies and derided their ability to help the war effort. In dire tones, the *Times* warned us that we were entering a meat grinder that had chewed up the Soviet Army beyond recognition.

Behind the headlines, the *Times* was using its polling to prove that the public was consumed with worries about military action and to make the point that public patience with a war in Afghanistan could wear thin quickly.

In a survey conducted between September 20 and 23 of 2001—less than two weeks after 9/11—the *Times* did its best to probe for weakness in public support for a war. In that survey, the *Times* first asked the vanilla question "Do you think the U.S. should take military action against whoever is responsible for the attacks [on 9/11] . . ." Predictably, 92 percent said yes.

But then the newspaper picked away at the result, wording its questions to try to shape the opinion of the respondents—a technique political survey researchers have called "push polling." The questions, each designed to raise doubts among the respondents, probed whether they would still support military action even if it meant:

". . . innocent people are killed?"

". . . going to war with a nation that is harboring those responsible for the attacks . . . ?"

". . . many thousands of innocent civilians are killed . . . ?"

". . . thousands of American military personnel will be killed?"

". . . the United States could be engaged in a war for many years . . . ?"

Wilting under the cascade of frightening possibilities, support for the war in the *Times* poll dropped from its initial level of 92 percent to 67 percent.

Finally, the *Times* biased the issue even further, posing a "when-did-you-stop-beating-your-wife" question: "In a war against terrorists and the country or countries that harbor them, how many American soldiers would you expect to lose their lives . . . ?" Then it biased the result by suggesting the answers: "Under one thousand, one thousand to five thousand, or more than five thousand?"

Voters split evenly, with about one-third giving each answer.

By listing the answers in the question, the *Times* obviously framed the respondents' choices. And the eventual results of the war prove the point: only *forty-seven* American soldiers were, in fact, killed in the war on Afghanistan. The magnitude of the *Times*'s distortion is evident.

An unbiased survey would certainly have probed for weaknesses in the prowar sentiment. But it would have alternated questions that raised doubts with those that tended to affirm military action. It would have inserted prowar questions in between the antiwar ones—asking, for example, what the respondents would think of going to war if there were hard proof that Afghanistan's regime was behind the 9/11 attacks, or if casualties could be held to a minimum, or if smart bombs reduced civilian deaths.

The fact that the war in Afghanistan was actually won so quickly makes the hypotheses tested in the *Times*'s survey seem far-fetched in retrospect. But at the time they did much to lead liberals to hesitate in backing the war.

More recently, in the war in Iraq, the *Times* also emphasized—day after day—the risks of the war, eagerly reporting any bad news, and downplaying the reality that the war was going very, very well.

In the last few months of 2001, the *Times* began laying out the other liberal themes it would hammer home endlessly in the year ahead. Chief among these were two new villains: the Bush administration's alleged intention to erode civil liberties protections and the purported growth in anti–Arab American bias among average Americans.

November–December 2001

The Times *Finds a Cause: The Civil Rights of Terror Suspects*

After the invasion of Afghanistan came off without major American combat casualties, and the natives of Kabul feted their American liberators, the momentum of the right seemed only to gather steam. The left desperately needed an issue with which to fight back. The *Times* served one up on a platter: civil liberties.

In the closing months of 2001, Raines's *Times* became almost hysterical in its warnings that the war on terror would lead to a massive effort to undermine the civil liberties of the average American. As the hundreds of captured Taliban terrorists—and a single American defector—streamed into the Guantanamo Naval Base in Cuba, the *Times* grew increasingly shrill about the civil liberties of the new prisoners. When the Bush administration proposed measures to prevent future terrorist attacks, the *Times* responded by zeroing in on the potential constitutional implications of the new investigations.

At a moment when we faced the most dramatic foreign attack on the continental United States in our history, the *Times* chose to focus on a largely phantom counterthreat of supposedly equal danger—the so-called erosion of civil liberties. Like a red scare injected to keep the public from recovering its true bearings, the *Times* pushed the idea that our liberties were under siege.

By November 2001, the newspaper was running almost daily articles, usually on page one, devoted to the civil rights and liberties of ordinary Americans, legal and illegal Arab immigrants, and prisoners captured by American armed forces. In the thirty-three days between November 9, 2001, and December 12, 2001, the *New York Times* had run twenty-nine front-page stories on these subjects, in its obsessive focus on civil rights and liberties.

Sounding the alarm on September 13, 2001, just two days after the attack, the paper ran an article by Clyde Haberman likening antiterrorism investigations to the killing of an African American on a stoop of a New York tenement by overzealous cops. The article, headlined "NYC: Diallo, Terrorism and Safety vs. Liberty," posited, "It is quite possible that America will have to decide, and fairly soon, how much license it wants to give to law enforcement agencies to stop ordinary people at airports and border crossings, to question them at perhaps irritating length about where they have been, where they are heading and what they intend to do once they get where they're going. It would probably surprise no one if ethnic profiling enters the equation to some degree." The author continued: "The prevailing ethic, certainly in post-Diallo New York City, is that profiling on the basis of race, religion, ethnic background and so on is inherently evil." While the article conceded that Americans *might* accept "intrusive law enforcement tactics" after 9/11, the implied metaphor—that the FBI's conduct could be likened to the trigger-happy police who shot Diallo—was as offensive as it was unfounded.

Day after day after 9/11, the newspaper devoted its front page to articles about civil liberties issues, forcing them to the forefront of the public debate:

- 9/13: "After the Attacks: Relations: Arabs and Muslims Steer Through an Unsettling Scrutiny"
- 9/16: "Broader Spy Powers Gaining Support"
- 9/19: "75 in Custody Following Terror Attack Can Be Held Indefinitely"
- 9/20: "Senate Democrat Opposes White House's Antiterrorism Plan and Proposes Alternatives"
- 9/23: "Americans Give In to Race Profiling, Once Appalled by the Practice, Many Say They Now Do It"
- 9/28: "In Patriotic Time, Dissent Is Muted"

Raines was flooding the zone.

The articles were often as one-sided as their screaming headlines suggested. In the September 20 story about the antiterrorism legislation, Philip Shenon and Neil A. Lewis reported Senator Patrick Leahy's opposition to the Bush administration's proposal to make it easier to detain and deport immigrants suspected of terrorism. (The bill allowed immigrants to be held for forty-eight hours and permitted indefinite detention of immigrants in a national emergency.) "We do not want the terrorists to win by having our basic protections taken away from us," Leahy intoned, while the paper reported the enthusiastic agreement of the American Civil Liberties Union (ACLU). Not a single supporter was quoted in the piece. Is it at all conceivable that Shenon and Lewis couldn't find one? Nine days after 9/11?

In the months after 9/11, the pounding on civil liberties issues continued. In one period, from November 9 to November 30, 2001, the *Times* ran sixteen front-page headlines in twenty-one days on civil liberties issues, most implying criticism of the Bush administration's antiterror policies.

Each time the administration would take a step to protect us from terrorism, the *Times* was there, sounding warning sirens about the possible encroachment of our liberties.

When the government scrutinized applications for visas by Arab men, the *Times* article headlined, "Longer Visa Waits for Arabs."

As FBI agents examined Arab men here on student visas—the same kind held by many of the 9/11 terrorists—the *Times* warned, "U.S. Has Covered 200 Campuses to Check Up on Mideast Students."

When Bush took executive action to stop new terrorist attacks, the

Times reported, "White House Push on Security Steps Bypasses Congress . . . Administration Urges Speed in Terror Fight, but Some See Constitutional Concern." In case anybody missed the point, the paper was back at it with a front-page story on November 18: "Civil Liberty vs. Security: Finding a Wartime Balance."

On November 21, the *Times* found something to cheer about when it reported Portland, Oregon's, refusal to cooperate with the FBI in questioning Middle Eastern male immigrants. The *Times* spared no effort to spread the Portland revolt. The next day, a front-page headline read: "Police Are Split on Questioning of Mideast Men, Some Chiefs Liken Plan to Racial Profiling." The articles, of course, gave short shrift to the thousands of cities whose police were cooperating fully with the Bureau in this reasonable step to strengthen its sources of information about terrorism.

The blasting continued—virtually every day's newspaper featured a blaring headline emphasizing the civil liberties issues raised by the Bush domestic antiterror crusade.

- "Bush's New Rules to Fight Terror Transform the Legal Landscape"
- "Tact Amid Rights Debate: Justice Dept Instructs Local Officials on the Questions That Should Be Asked"
- "Al Qaeda Link Seen in Only a Handful of 1200 Detainees"
- "Groups Gird for Long Legal Fight on New Bush Anti-Terror Powers"
- "Religious and Political Groups in U.S. Could Again Be Fair Game Under New Plan"
- "Few in Congress Questioning President Over Civil Liberties"

Reading the *New York Times* during November and December of 2001, one could easily get the impression that the central problem facing America wasn't so much terrorism as the erosion of our liberties.

Even when, on December 12, the *Times* announced the arrest of Zacarias Moussaoui, the twentieth hijacker who was unable to participate in the 9/11 attacks because he had been arrested the month before, it found a way to put a liberal spin on the story. The *Times* headline read: "Man Held Since August Is Charged with a Role in Sept 11 Terror Plot." Apparently, the key fact was that Moussaoui's rights had been violated by the long imprisonment—not that he was in on the plot that had killed three thousand people.

And even when a suspect was released, the *Times* was equally critical: "Cleared After Terror Sweep, Trying to Get His Life Back."

Did the issue merit such intense focus? Were our treasured civil liberties really that much in danger during this period?

Clearly, the general public never shared Raines's and the *New York Times*'s apoplectic view of the issue. Even the newspaper's own poll of December 7–10, taken as the paper's intense coverage of the civil liberties issue was winding down, confirms as much.

Like a politician tracking his campaign to see what impact it is having, the *Times* asked Americans what they thought—in the midst of its own near-daily pounding of front-page stories about civil liberties. Thirty-one percent said they had heard "a lot" about the way the Bush administration was trying to "seek, investigate, and prosecute suspected criminals." But only 12 percent thought the administration was going too far. Were Americans worried "that some of these changes may apply to people like you"? Only 8 percent were "very worried."

Finally, asked if respondents were concerned about "losing some of [their] civil liberties," only one-third said they were. Only one-quarter felt the government was currently violating their constitutional rights (and many of those complaints were about gun ownership limitations and antismoking rules). How disappointed the liberal editors must have been!

Still, like a candidate trying to put the best possible face on bad poll numbers, the *Times* nevertheless claimed to find evidence in its poll of public doubts about the antiterror measures. The headline it gave to its own disappointing poll: "Public Is Wary but Supportive on Rights Curbs." The article noted that while "Americans are willing to grant the government wide latitude in pursuing suspected terrorists," they "are wary of some of the Bush administration's recent counterterrorism proposals and worried about the potential impact on civil liberties."

The results must have been especially disappointing, given the *Times*'s efforts to slant the survey. Incredibly, the newspaper put respondents through a series of thirty-three questions, examining each aspect of the civil liberties issue; the entire survey would take more than fifteen minutes to complete. Naturally, as the respondents heard question after question on the same subject, their sensitivity to the constitutional issues was heightened. As any experienced pollster knows, cooperative respondents often grow anxious to please the pollster; increasingly alarmed by the scenarios described in the survey, many likely gave the answers the questioner clearly wanted to hear.

The *Times*'s survey also asked questions in an order bound to prejudice the sample and encourage it to give the answer the *Times* wanted. For example, the questions noted that "in the past" suspects have been tried in "criminal court requiring a jury, a unanimous verdict, and a civilian judge." Was this the right way "to deal with suspected terrorists"? By a margin of 53–37, the sample gave the answer the *Times* wanted to hear: *yes*.

Would you rather try terrorists "in open criminal court, with a jury, a unanimous verdict, and a civilian judge," or "in a secret military court, with a military judge and without a unanimous verdict?" Asked that way, the polling sample voted 50–40 for the open procedure.

An unbiased poll might have mentioned to respondents that the president and the attorney general felt that the evidence adduced against terrorists at open trials might aid al Qaeda and provide them with vital information about our defenses. But the *Times* wanted to make the case for a liberal response to Bush's proposals.

Then the questions turned to the civil liberties of Arab Americans. Should the government be "allowed to investigate religious groups that gather at mosques, churches, or synagogues without evidence that someone in the group has broken the law or does that violate people's rights?"

The newspaper failed to mention who, precisely, was proposing to investigate churches or synagogues? How many Arab American terrorists did the *Times* think would be holed up in the Church of Our Blessed Lady or Temple Rodolf Sholum? The *Times,* obviously, threw in the idea of non-Islamic religious facilities to get the answer it wanted. And it did: 75 percent said they opposed the practice.

Finally, having asked about every possible violation of civil liberties, the *Times* closed in for the kill. "Which concerns you more right now . . . that the government will fail to enact strong antiterrorism laws or that the government will enact new antiterrorism laws which excessively restrict the average person's civil liberties?" After the battery of questions highlighting civil liberties, the *Times* managed to get 45 percent to say they were more concerned about civil liberties, while 43 percent were more worried about terror.

After the long, long series of questions, seemingly bending and straining to justify the newspaper's editorial bias, the *Times*—barely— got the result it wanted: cause to report that Americans were "wary" about their civil liberties.

With the hindsight of history, how justified was the *Times* in its almost monomaniacal focus on civil liberties and the rights of Arab Americans? In fact, there were relatively few incidents of anti–Arab American outbursts following 9/11. No mosques were burned, no Arab Americans lynched. Hate crimes against them rose, unfortunately, but only 481 were reported for all of 2001—a tiny proportion of the 3 million Arab Americans living in the United States.

But the real admission that the *Times* was overreacting came in the form of the paper's own behavior: After its lukewarm December poll, it cut way back on its coverage of civil liberties issues. In the nine months from February to October 2002, the *Times* ran fewer than one front-page story per week on civil liberties issues (only twenty-seven over the nine-month period). It would seem that even the *Times* didn't think, in retrospect, that the intensity of its coverage had been justified.

Editorial Support for the Florida Professor with Suspected Terrorist Links

Raines's "flood the zone" policy does not apply to all the news that's fit to print at the *Times*. To the contrary, if a story doesn't fit its liberal bent, there's *no* pounding, *no* flooding, *no* repetition. And if a liberal story turns bad, there's no correction or admission from the liberal editors. A good example is the tale of a Florida professor, Sami (short for Osama) Al-Arian.

The professor first came under fire from Bill O'Reilly, my gutsy colleague at the Fox News Channel, who invited Al-Arian to be a guest on his show on September 26, 2001, two weeks after 9/11. For almost a decade, the FBI had believed that Al-Arian was tied to violent terrorist groups. Never one to duck a controversial issue, O'Reilly challenged

Al-Arian by playing an old video clip in which the professor called for "death to Israel." O'Reilly bluntly suggested that the CIA should be following Al-Arian at all times.

After the show, the University of South Florida received hundreds of phone calls and letters complaining about Al-Arian, as well as several death threats. The university suspended him for failing to make it clear that his extremist views were not those of the university and for creating a dangerous environment on campus.

Al-Arian immediately became a cause célèbre among liberals and academics, who touted his First Amendment right to speak and screamed about his right to "academic freedom."

At first, the Al-Arian story seemed to have big "flood-the-zone" potential for the *Times*. On January 27, 2002, the paper's editorial page forcefully attacked Al-Arian's sacking in, "Protecting Free Speech on Campus," which claimed that his termination by the university made "a mockery of free speech."

Almost a month later, *New York Times* columnist and Pulitzer Prize–winning journalist Nicholas Kristof took up the Al-Arian mantra. In a column entitled "Putting Us to the Test," on March 1, 2002, Kristof warned of the dangers of a society that silences dissent in times of difficulty. Kristof made no mention of a finding by a federal judge in 1997 that there was credible evidence that Al-Arian's Muslim group did, in fact, have ties to terrorist organizations. Instead, Kristof described the now-indicted terrorist as a cuddly patriot:

> a rumpled academic with a salt and pepper beard who is harshly critical of Israel (and also of repressive Arab countries)—but who also denounces terrorism, promotes inter-faith services with Jews and Christians, and led students at his Islamic school to a memorial service after 9/11 where they all sang *God Bless America*.

God help us.

As it turned out, O'Reilly was right after all.

According to an article by David Tell in the *Weekly Standard* of March 15, 2002, the FBI and the U.S. Attorney's office in central Florida had confirmed about a week before Kristof's column that there was an ongoing investigation of Al-Arian. Kristof was so focused on

Al-Arian's First Amendment rights that he ignored his invocation of the Fifth Amendment nearly one hundred times when asked, in August 2001, whether he had provided financial assistance to terrorists.

Kristof and the *Times* did not really address the whole story of Al-Arian and his leadership in one of the most violent terrorist groups—right on American soil. In fact, about five weeks later, on April 16, 2002, Kristof wrote a column entitled "Behind the Rage," in which he quoted Sami Al-Arian as saying: "The Israeli occupation represents a total humiliation of all the Arab regimes."

Kristof never mentioned any controversy about his new best source. Instead, he was presented simply as a "Palestinian activist."

Then the FBI pounced, arresting Al-Arian on charges of conspiracy to commit murder and aiding and financing the violent terrorist group. According to the indictment, Al-Arian and others used the academic environment of the University of South Florida as a cover for vicious terrorists.

Did the *Times* backtrack on its earlier stories? No way. Any corrections? Apologies? New editorials? Nope.

What had started as a flood-the-zone story was now barely a trickle. In fact, it was a drought.

January–February 2002

Can the Enron and Wall Street Scandals Save the Democratic Party?

Seeing that the civil liberties and free speech issues weren't cutting it with the American voters, the *Times* seized on a new issue when the Enron scandal burst, drenching the stock market and the economy. Raines was off and running with high-pitched coverage from day one. Unable to breech the wall of Republican/Bush popularity by flooding the zone with civil liberties, the *Times* shifted its focus, replacing intense coverage of constitutional rights with a preoccupation with the Enron scandal and the rash of Wall Street debacles that followed.

Between January and April, in a desperate effort to lead the nation away from a single-minded focus on terrorism, the newspaper put 125 articles on its front page about the Wall Street scandals—almost twice as many as it ran about the war on terror and the battles in Afghanistan *combined*. In January alone, the paper ran fifty-one front-page stories on the scandal. (Meanwhile, civil liberties were dwindling as an issue; the *Times* ran only eight front-page stories about it that month.)

From the beginning, the *New York Times* saw the Enron scandal as a political story. Apparently convinced that the Bush administration must have granted special favors to a dying Enron, and determined to link them to campaign contributions to the president, the *Times* played the scandal, from the start, as a partisan story. Raines kicked it off on January 5 with the headline: "Democrat Assails Bush on Economy." Then, narrowing its focus to Enron, the *Times* reported on January 11 that "Enron Contacted Two Cabinet Officers." The next day's story blared: "Enron Sought Aid of Treasury Department to Get Bank Loans." Not mentioned in the headlines was the fact that the official doing the asking was former *Clinton* Treasury secretary Robert Rubin.

Anxious to cast Enron as a domestic political story, the paper teed the issue up for the Democrats on January 12 with the headline, "Parties Weigh Political Price of Enron's Fall." Two weeks later, the *Times* answered its own question, reporting that their "Poll Finds Enron's Taint Clings More to G.O.P Than Democrats; Economy, Not Terror, Is Now Perceived as Highest Priority."

The *Times* story described what it said were the poll's findings: "Americans perceive Republicans," the story ran, "as far more entangled in the Enron debacle than Democrats and their suspicions are growing that the Bush Administration is hiding something or lying about its own dealings with the Enron Corporation."

Really?

Not if you look closely at the data. The actual question asked in the survey was: "When it comes to their dealings with Enron executives prior to Enron's bankruptcy, do you think members of the Bush administration are telling the entire truth, are mostly telling the truth but hiding something, or are they mostly lying?" Only 9 percent said the Bush administration was lying. Seventeen percent felt they were being entirely truthful, and an additional 58 percent said they were "mostly telling the truth but hiding something." So the poll actually found that 75 percent (58 percent + 17 percent) of the voters felt Bush was mainly being truthful—quite the opposite of the *Times*'s spin that "a majority of Americans say the Bush administration is either hiding something or lying about Enron."

The article clearly left the impression that Bush was having a tough ride over Enron. But the questions cited by the *Times* actually addressed perceptions of the Bush *administration*. President Bush himself came through the poll with flying colors. Asked if Bush cared about "the needs and problems of people like you," voters agreed by 76–23 that he did. Asked how they rated Bush's handling of the economy, voters backed it by 56–33. But you'd never know all that from the *Times*'s coverage: Though the paper did note that the president had high approval ratings, it spent most of its time reporting on the administration's purported credibility gap.

But Raines's strategy of flooding the zone with articles about Enron didn't do the trick. It became clear, as the investigation proceeded, that Bush's people had done nothing to help Enron avert bankruptcy, despite the requests from Democrat Robert Rubin. The Bush Justice Department promptly indicted top corporate officials at Enron and other companies and closed down the Arthur Andersen accounting firm by indicting it as well.

As corporate executives were led away in handcuffs, Bush's job-approval ratings remained very high, seemingly unaffected by the story. Democratic chances in the midterm election looked pretty bad. The constant stream of articles about Enron and corporate scandal had done a lot to reduce stock prices, but not much to cut Bush's approval ratings.

The *Times* needed a new strategy. The relentless suicide/homicide bombings of Israel—amounting to the Holocaust on an installment plan—provided it. The newspaper became determined to push Bush into a new campaign of diplomatic activity, declaring that only by resolving the intractable Arab-Israeli mess could the administration attract the moderate Arab support it would need to go after Saddam Hussein in Iraq.

March–May 2002

The Times *Puts the Palestinian Problem Ahead of Invading Iraq*

From the very beginning of the war on terror, the *New York Times* had pushed the line that Bush needed to get more personally involved in resolving the dispute between Palestinian terrorists and Israel. The paper saw U.S. policy on Israel as directly linked to the 9/11 attacks.

In the days after 9/11, the newspaper asked voters in its survey whether they "blamed" the attacks on U.S. polices in the Middle East over the years. What a choice of words! "Blamed" implies not just causality but fault as well. Here's how the question read:

> When you think about the terrorist attacks on the World Trade Center and the Pentagon, do you place a lot of blame, some blame or no blame at all on United States policies in the Middle East over the years?

The responses were as follows:

A lot	14%
Some	54%
No blame at all	25%
Don't know	7%

(September 20–23, 2001, Going to War? #30)

What the use of the word "blame" seems to have reflected best was not the views of the voters, but those of the *Times* itself. The newspaper suddenly seemed stuck on the idea that Bush must do more to wind up the decades-long battle between the PLO—and its terrorist allies Hamas and Hezbollah—and the Israeli government, before he could move to dethrone Saddam Hussein in Iraq or to take the war on terror much further. As the suicide/homicide bombings increased, the *Times* grew more and more insistent.

In editorials and op-eds, the newspaper pushed the connection. Here's a sample:

From a news analysis by David E Sanger, April 1, 2002:

> To build Arab support for his impending confrontation with Iraq, Mr. Bush knows he cannot afford to alienate other Arab nations, whose anti-Israel declarations have grown in vehemence and urgency, along with their demands that Mr. Bush restrain the Sharon government . . . [some conservatives say that] unless Mr. Bush convincingly demonstrates his desire to act as peacemaker, and show the way both to the creation of a Palestinian state and a secure Israel, he cannot move against Iraq at all.

From a news analysis by Patrick E. Tyler, April 18, 2002:

> The prospect of more violence in the Mideast served to undermine the administration's efforts to build a new coalition of Arab states that might support military operations against Iraq later this year.

From an op-ed by Thomas L. Friedman, June 9, 2002:

> The State Department argues that ending the Israeli-Palestinian conflict is vital for turning back the anti-American tide in the Arab-Muslim world, for preparing the groundwork for any attack on Iraq and for securing Israel's long-term future.

From a news analysis by David E. Sanger, April 4, 2002:

> The anti-American protests in several Arab capitals have also caused concern within the administration that the conflict could spread instability across the region and undercut the possibility of an American military campaign against Iraq. [A letter from conservatives] also urged Mr. Bush to "accelerate plans for removing Saddam Hussein from power in Iraq." Middle East experts, including some in the State Department, say a confrontation with Iraq now would further inflame anti-American protests and force American allies to split from Washington.

It was as if the *Times* were pushing Bush into quicksand—urging his greater involvement in the Middle East, knowing full well how this tar pit had swallowed up presidents and secretaries of state for decades.

Bush was clearly thrown off balance by the growing crescendo demanding that he get more involved in Middle East diplomacy. Without the clarity of his good vs. evil positioning in the war on terror to guide him, the president seemed to lurch between condemning Israel and attacking Arafat and the PLO.

Buffeted by events—and by the constant urging of the *Times* for greater involvement and even-handedness in the Middle East—Bush began 2002 by considering a peace plan proposed by Saudi Arabia, calling for Israeli withdrawal to pre-1967 boundaries in return for recognition and promises of nonaggression by the Arabs.

In March 2002, Bush rebuked Israel for its use of military force to fight terrorists in the West Bank. The next month, he sent Secretary of State Colin Powell to the region to attempt to negotiate a settlement. The Powell trip, hailed by the *Times,* represented exactly the sort of focus on the region the newspaper had been calling for. But, like other secretaries of state, Powell failed to pull a rabbit out of his hat; the bloodshed only intensified.

Bush looked like a man adrift. But in the weeks after Powell's return, the president seemed to grasp that the war on terror he had been waging in Afghanistan was the same battle Israel had been fighting on the West Bank. Calling for the removal of PLO leader Yasser Arafat, Bush came down more firmly on the side of Israel. In the process, the president regained his political footing, and his period of dithering and hesitation appeared to be over. He had emerged from the quicksand.

As Bush attempted to find his way through the Middle East labyrinth, the *Times*'s coverage of the Israeli-Palestinian conflict was, itself, decidedly pro-Palestinian. The paper featured a focus on the families of suicide bombers, while giving shorter shrift to the orphans and widows of their victims.

Between January and June 2002, the *Times* ran thirty-eight front-page articles with headlines alluding to Palestinian deaths or Israeli military action, compared with only twenty-six about Palestinian suicide/homicide bombers attacking Jews.

But it was in the paper's feature articles and color commentary that the *Times*'s bias became most evident. Between March and June, only three front-page articles covered Israeli pain and anguish as a result of the conflict, compared with twelve about Palestinians.

As Israel retaliated against the homicide/suicide bombings that killed its citizens, the *Times* kept a close chronicle of Palestinian suffering:

- April 3, 2002: a front-page headline told readers that "Anger in the Streets Is Exerting Pressure on Arab Moderates"
- April 4: "Arabs' Grief in Bethlehem, Bombers' Gloating in Gaza"
- April 8: "In Nablus's Casbah, Israel Tightens the Noose"
- April 11: "Attacks Turn Palestinian Plans into Bent Metal and Piles of Dust"
- April 13: "Jenin Refugee Camp's Dead Can't Be Counted or Claimed"

- April 14: "Refugee Camp Is a Scene of Vast Devastation"
- April 16: "For Palestinian Refugees, Dream of Return Endures"
- April 21: "In Rubble of a Refugee Camp, Bitter Lessons for 2 Enemies"

But, beyond the issues of bias, the very format of tit-for-tat coverage belied the basic lack of equivalency between those who perpetrated the terror in Israel and the Israelis who responded in self-defense. As the headlines ping-ponged back and forth, the cause/effect relationship of Arab provocations and Israeli reaction was lost.

Particularly galling was a series of *Times* stories that seemed to help enshrine in martyrdom the suicide/homicide bombers who preyed on innocent Israelis. On June 30, 2002, Elizabeth Rubin wrote an eight-thousand-word portrait for the Sunday *Times Magazine* about Qeis Adwan, a much-wanted Palestinian terrorist slain by Israeli forces on April 5, 2002.

Noting with splendid disregard for cause and effect, "By now, Israeli assassination operations against Palestinians have become as routine as Palestinian suicide bombings," Ms. Rubin painted a compelling portrait of a misunderstood but idealistic young man. The article even quoted the terrorist's mother, telling us that he "never carried a gun," "was an angel in a human body," and "didn't even like to see insects die" when he was a boy.

Apparently he changed his mind. The article noted that an Israeli army officer cited his "ability to manufacture ever more potent bombs [and] his logistical imagination in the plotting and execution of attacks . . ." as reasons for his celebrity among Israeli police.

Noting that Adwan was "the most popular and inspiring leader of the student union" at college, the article briefly alluded to the curious fact that this young man had sent a suicide bomber to a restaurant, carrying his most creative bomb design, killing fifteen and wounding more than forty when Adwan's creation exploded.

The article reported that his friends felt he was " 'kind,' 'simple,' 'flexible,' 'polite,' 'diligent,' and 'beloved,' " and it stressed his commitment to architecture as a future occupation. (Presumably after he had exhausted his interest in demolition.)

Mesmerized by her cute little bomber, Ms. Rubin tells us, "Everywhere I went on campus, I heard stories of Qeis' efforts to solve students' problems. 'He found my sister housing and lowered her tuition fees,' a local journalist said. . . . He was the poor students' advocate, collecting funds from rich families to give to the poor. . . . Students of every political persuasion sought him out for help with their psychological, financial, and academic problems."

Reporting on his two previous arrests, Ms. Rubin takes care to inform us that he was "on affable terms with everyone, even his jailers."

Such adulatory coverage of these terrorist thugs was only the most blatant of the examples of the *Times*'s bias.

Frequently, the paper would run articles featuring an Israeli attack on Palestinians, emphasizing the casualties and the crowds mourning the victims—but burying the reason for the attack. On July 24, 2002, for example, the paper ran the headline "Gaza Mourns Bombing Victims; Israel Hastens to Explain."

The accompanying story began by describing the "tens of thousands of mourners streaming in a three-mile procession . . . through Gaza City's bleak streets for the funeral of the Hamas military chieftain, Sheik Salah Shehada." The article went on to describe how Sheik Shehada was killed when an Israeli plane dropped a bomb into "a densely packed neighborhood," leveling half a city block and killing 14 people and injuring 140 more.

The article quoted Israeli prime minister Sharon as calling the attack "one of our major successes," but noted that after the "worldwide condemnation began pouring in" (including a condemnation by Bush), a senior Israeli military official said, "We wouldn't have done it if we knew what the consequences would be." The story went on to detail the agony of the survivors, with accounts of the terror of the attack's suddenness.

Aside from a brief note that he was responsible for a string of deadly attacks, it wasn't until the twenty-fourth paragraph of the article, more than three-quarters of the way into the story and just before its end, that the *Times* bothered to explain why the Israelis so wanted to kill the Hamas leader.

Their target, as it happened, was the commander of the military wing of Hamas—according to the Israelis, "the most brutal and brilliant terrorist operating in the Gaza Strip . . . personally responsible for orchestrating attacks against hundreds of civilians over the past two years." Among the little pranks that he was planning or had up his sleeve were "placing a truck bomb under a bridge . . ." to kill Israeli settlers; "sending a boat filled with explosives to a bathing beach, and dispatching suicide bombers to a shopping mall and gunmen to attack a Gaza settlement."

No one can or should fault the *Times* for covering the anguish of the innocents killed by the Israeli attack. But any fair reader will find a great deal of fault in the newspaper's refusal to explain the reason for the attack in the lead of the story.

But then, as the spring of 2002 waned, so did the Israeli-Palestinian conflict, tapering off as unilateral Israeli military measures cooled Arab terrorism, if not Palestinian passions. No longer did it seem that Bush needed a deal on the West Bank before he could move against Iraq after all. So the *Times* shifted gears, yet again—this time back to the economy.

July 2002

The Times *Gives a Scandal . . . But Nobody Comes*

Desperate days among Democrats. Nothing had worked. The war in Afghanistan had been won, at minimal cost. Civil liberties had fallen flat as an issue. The Enron scandal seemed to have no political traction.

Like a drowning liberal, Raines tried to make a latter-day Watergate of the accusations of corporate malfeasance surrounding Bush's and Cheney's time in the private sector. With five front-page articles between July 8 and July 18, 2002, the *Times* worked to hang a black cloud to surround the president and vice president as the election approached.

Using the breathless lingo of scandal, the *Times* reported that Bush had been forced "on the defensive by questions about his role in a stock sale a dozen years ago."

When Bush defended himself against the makeshift scandal at a press conference on July 9, 2002, the *Times*—in its *news* coverage, not its opinion page—called the president's answers "vague and dismissive" and the timing of his press conference "curious." Saying that Bush was "in for a rude shock" if he had hoped to turn attention away from his

corporate dealings at his press conference, the paper reported that the
president "for most of the 36 minute session with reporters . . . faced a
series of detailed questions about his own business dealings . . . at
Harken Energy . . ."

The "scandal" arose because Bush had sold 212,140 shares of stock
in the Harken Corporation at $4 per share while he was in the private
sector, eight days before the company finished its second-quarter post-
ing with a loss of $23 million. One month after Bush sold the stock, it
was down to $2.37. The *Times* reported: "The S.E.C. investigated on
suspicions that the transaction was conducted on the basis of insider
information" but noted that the investigation was subsequently
dropped.

When Bush addressed a Wall Street gathering, on July 10, 2002, to
propose an overhaul in corporate regulation, the *Times* mentioned the
substance of the president's speech only far down in its front-page article. The headline and lead of the story attacked the president for receiving loans from Harken. "President Bush received two low-interest loans
to buy stock from an oil company where he served as a board member
in the late 1980s," the *Times* story read. "He then benefited from the
company's relaxation of the terms of one loan in 1989 as he was
engaged in the most important business deal of his career."

The *Times* mentioned Bush's proposals only by reporting, in the
same story, that "Mr. Bush called for a halt to those types of insider
transactions, challenging corporate directors to 'put an end to all company loans to corporate officers.'"

To see whether the public was angered by Bush's financial dealings,
the *Times* polled on June 18—only to find that Americans generally
trusted their president in his personal business dealings prior to taking
office.

Nevertheless, seemingly determined to salvage something from their
week of excoriating the president, the *Times* led its article with the
headline "Poll Finds Concerns That Bush Is Overly Influenced by Business." The article said, "The survey suggests that the unfolding revelations about corporate misconduct and inflated earnings hold
considerable peril for the White House and Mr. Bush's party in this
Congressional election year."

Far down in the story, the paper reported that Bush's approval rating stood at 70 percent; farther down, it noted that by 43–21 voters felt

Bush had "behaved honestly and ethically in his business practices while in the corporate world."

Summer of 2002

It's the Economy, Again . . . Stupid

As the midterm congressional elections approached, it became clear to Democrats that the opposition needed a big issue to beat Bush and to wipe away the credit the president was getting for fighting the war on terror. Raines had likely hoped that the Wall Street scandals would divert the national focus from Bush's antiterror crusade, but it hadn't worked out that way. Now, the paper went into overdrive in its effort to steer national attention away from terror and to focus it on the economy—the issue that the Democrats had chosen to use in defeating the Republicans.

Laying the groundwork for its campaign, the *Times* polled Americans on July 13–July 16, 2002, about the economy, corporate greed, the Wall Street accounting scandal, and the Bush-Cheney corporate past at the Halliburton Corporation and at Harken Energy.

Incredibly, the survey had hardly a single question—and the ensuing articles covering it hardly a single word—about terrorism, Iraq, threats to the United States, the Middle East, Afghanistan, al Qaeda, or any other such topic. Less than a year after the Twin Towers in New York City had collapsed, six months after the war in Afghanistan, and only a few weeks before the showdown with Iraq escalated, the newspaper was opting to ignore these subjects altogether. It was as if the *New York Times* was sending a signal to its readers and the nation's opinion leaders: Get off the war on terror and focus on the economy. It's the way for the Democrats to win.

Deliberately or not, this survey was one of the most heavily weighted that the newspaper conducted during the year. When the original sample turned up 31.8 percent Democrats and 31.0 percent Republicans, the *Times* weighted the numbers to produce a 34.2 percent/28.4 percent Democratic edge in the survey sample. This five-point "correction," of course, had a very direct impact on the data and the resulting conclusions that the *Times* published as fact.

Still, despite weighting the sample, the newspaper's thesis that the national agenda had shifted away from terror and toward the economy ran into difficulty from the very start of the survey. The *Times* began by asking, "What do you think is the single most important problem for the government—that is, the president and Congress—to address in the coming year?" Obstinately, 22 percent said "terrorism/defense/war" was uppermost in their minds; 19 percent said "economy/stock market/loss of jobs."

And even when the July survey found extensive public concern about the economy and anger against corporate malfeasance at Enron, WorldCom, Arthur Andersen, and their cornucopia of white-collar criminals, it also found a decided reluctance to tie their misdeeds to President Bush.

While respondents broke evenly, 49–49, on whether the economy was good or bad, they gave Bush a 52–37 approval rating on the way he was handling it.

The pattern prevailed throughout—voters recognized the problems but liked Bush nevertheless. Voters agreed by 58–28 that big business had "too much influence" over Bush and divided evenly, 42–42, on whether Bush was "more interested in protecting the interests of ordinary Americans or the interests of large corporations." But they also insisted that Bush "cares about the needs and problems of people like yourself" by 68–27 and agreed by 80–14 that he "shares the moral values most Americans try to live by."

The *Times*, of course, accentuated the negative ratings and buried the positive ones in its news coverage of the poll.

And in doing so, it sent an unmistakable signal to the Democratic Party: "The survey suggests that the unfolding revelations about corporate misconduct and inflated earnings hold considerable peril for the White House and Mr. Bush's party in this Congressional election year."

Now the *Times* decided to try a new variant of the Raines strategy. It didn't flood the zone, it parched it. Suddenly, the *Times*'s coverage of terrorism virtually ground to a halt. In July and August of 2002, the *Times* ran only thirteen articles on its front pages relating either to Afghanistan or to the challenges of the war on terror (excluding Iraq and the Israeli-Palestinian conflict). Of the thirteen, four related to Afghan or Paki civilian deaths, and three others focused on American attempts to keep its peacekeeping forces immune from jurisdiction by the new World Criminal Court. Only six articles in sixty-two days ran on the front page related specifically to the global war on terror.

Instead, it was the economy, stupid.

August–September 2002

Against War in Iraq

But terror would not go away as a political issue. Despite the newspaper's best efforts to portray to Americans that the economy was the paramount issue, terror kept reasserting itself as the central national concern—driven by alerts, warnings, terrorist arrests, and national apprehension.

So, in the late summer of 2002, the *Times* seemed to light on a new idea to sell: the notion that Americans felt Bush was *losing* the war on terror.

Basing its comments on a survey timed to coincide with the first anniversary of the 9/11 attacks, the *Times* reported that "Americans increasingly doubt that their government has done enough to protect them against terrorist attacks. . . ."

But respondents to the *Times's* poll said nothing of the sort. As usual, the newspaper was reading the data to get the answers it wanted.

The survey, in fact, reflected deep and broad national confidence in Bush's leadership and the results he was achieving. By 76–22, voters expressed "a great deal" or a "fair amount" of "confidence in the ability of the U.S. government to protect its citizens from future terrorist attacks," hardly the resignation to future attacks the *Times* professed to see in the data. Even though 68 percent of the American people approved of Bush's handling of terrorism, George W. couldn't satisfy the *New York Times*.

So how did the *Times* manage to draw a negative conclusion from the survey? As usual, by interpreting the data in light of its own opinions. In answering each question about how much progress the administration was making in the war on terror, respondents could give one of four answers: "a lot of progress," "some progress," "not much progress," or "no progress at all."

The normal way to read the data would be to combine the two positive responses (that "a lot" of progress was being made and "some" progress was happening), and then combine the two negative responses (that there was "not much progress" or "no progress at all"). When you combined the two positives against the two negatives, the result showed a broad public approval of the effectiveness of Bush's efforts:

- 83 percent saw progress in "closing terror training camps in Afghanistan"
- 66 percent saw progress in "eliminating the threat from terrorists operating from other countries"
- 55 percent saw progress in "putting a stable government in place in Afghanistan"
- 80 percent saw progress in "developing a comprehensive plan for protecting the U.S. against terrorism"
- 82 percent saw progress "in making air travel safe"

In fact, it was only in "improving the image of the United States in the Arab world" that less than a majority cited progress (only 40 percent saw progress in this area).

But that wasn't how the *Times* presented the data. It reported as positive only those who said there was "a lot of progress" and lumped those who said there had been "some progress" in with those who saw "not much progress" or "no progress at all." By combining three less-than-fully positive responses and branding them all as negative, the *Times* found a way to report that voters were unimpressed by Bush's conduct of the war on terror. Talk about fuzzy math.

The *Times*'s pattern continues: slant the poll by weighting the sample, stack the poll by the choice of questions, and then misinterpret it to ignore most of the pro-Bush responses.

But no matter how the newspaper misinterpreted its own polling, it was evident that Bush was escaping blame for a bad economy and winning plaudits for the war on terror.

So the *Times* seemed to open a second front in its war on Bush, ratcheting up its opposition to waging war in Iraq against the dictatorship of Saddam Hussein. Having largely failed to distract attention from the war on terror by its obsessive focus on the economy, the *Times* began to push the view that Bush was rushing to war in Iraq, ignoring the need for consultation with our allies, Congress, and the United Nations.

Throughout August, September, and October, the newspaper delivered its relentless party line, using headlines, editorials, placement of stories, and, above all, its surveys to make its case.

Almost every day's front page brought another article speculating on the possible adverse consequences of a war with Iraq, while few focused on the possibility of Iraqi terror attacks against the United States if we did nothing.

The *Times* warned that the war might push the economy into recession:

- "Profound Effect on U.S. Economy Is Seen from a War Against Iraq"

It could lead to major American combat casualties:

- "Air Power Alone Can't Defeat Iraq, Rumsfeld Asserts. Cites Secret Mobile Labs"
- "Iraq Said to Plan Tangling the U.S. in Street Fighting; Can't Win Open Battles. Hoping to Frighten Washington by Raising Political Specter of House-to-House War"

Washington might be left without allies:

- "Kurds, Secure in North Iraq Zone, Are Wary About a U.S. Offensive"
- "Administration Seeking to Build Support in Congress on Iraq Issue; France and Britain Press for Working with U.N."
- "Bush Asks Leaders in 3 Key Nations for Iraq Support; Little Headway Apparent"
- "Rift Seen at U.N. over Next Steps to Deal with Iraq; Bush Asks Tough Action; U.S. Distrusts Inspections, but Other Nations Want to Give Them a Try Before War"
- "U.S. Hurries; World Waits; Bush Left Scrambling to Press Case on Iraq"

Congress, including even Republican members, were opposed to Bush's policy:

- "Call in Congress for Full Airing of Iraq Policy"
- "Top Republicans Break with Bush on Iraq Strategy; Cites Risk of a War Plan; Current and Former Foreign Policy Figures Urge More Diplomatic Preparation"
- "President Notes Dissent on Iraq; Vowing to Listen—'Healthy Debate' on War in Quick Answer to Republican Concerns, He Reserves Sole Right to Decide Course"
- "Bush to Put Case for Action in Iraq to Key Lawmakers; Senators Not Convinced Powell Speaks of Differences in the Administration over Dealing with Hussein"
- "In Senate, a Call for Answers and a Warning on the Future; Focus on Iraq Criticized"
- "Democrats, Wary of War in Iraq, Also Worry About Battling Bush"

And, anyway, our own hands aren't exactly clean:

- "Officers Say U.S. Aided Iraq in War Despite Use of Gas—Battle Planning on Iran—New Details of 1980's Program—Help Continued as Iraqis Used Chemical Agents"

Bush may have used psych-ops to fight the war in Iraq, but the *New York Times* was conducting its very own psych-ops campaign to prevent it.

By raising the specter of street fighting in Baghdad, bringing up the possibility that the United States conspired with Iraq in its war with Iran, focusing on dissent over the war, and predicting dire economic consequences from a conflict, the *Times* hyped the antiwar effort.

In September and early October, the *Times* pressed Bush to submit his case for invading Iraq, first to Congress and then to the United Nations. Hoping to slow down what the paper saw as the administration's rush to war, the newspaper claimed that its surveys showed that Americans shared its point of view. Ignoring the evident widespread support for the Bush position, it chipped away through loaded questions to try to build a case against immediate American action.

Its surveys, and the way they were reported in the newspaper, exerted a great influence on opinion leaders, rallying top Democrats to an antiwar stand. In doing so, the newspaper likely misled American liberals into making the war the key issue in September 2002—with disastrous results for the Democrats in the midterm elections two months later.

"Poll Finds Unease on Terror Fight and Concerns About War on Iraq," screamed the front-page headline reporting the results of a survey the *Times* conducted on September 2–September 5, 2002. But once again, the poll data and the headline in the *Times* describing it were two different things. While the headline suggested public opposition to a war with Iraq, the poll itself found that most Americans backed military action.

Asked if they "approve or disapprove of the United States taking military action against Iraq to try and remove Saddam Hussein from power," respondents heartily endorsed the war by 68–24. Even Democrats agreed, by 59–33.

The poll also found that 79 percent felt that "Iraq currently possesses weapons of mass destruction" and that 62 percent agreed that it "is planning to use those weapons against the United States.

But the *Times* ignored this fair reading of public sentiment. Instead it based its headline on two loaded questions. The first asked:

Which statement do you agree with more?

A. Iraq presents such a clear danger to American interests that the United States needs to act now, even without the support of its allies [27 percent], or

B. The United States needs to wait for its allies before taking any action against Iraq [67 percent].

But this left the obvious question: What is Great Britain if not our "ally"? Led by its courageous prime minister, Tony Blair, the British were indeed backing military action against Iraq—right now. And, incidentally, so were Spain, Portugal, Holland, Italy, Poland, the Czech Republic, Hungary, Turkey, Romania, Bulgaria, and Denmark.

The second question was just as biased: "Should the United States take military action against Iraq fairly soon [35 percent] or should the U.S. wait and give the United Nations more time to get weapons inspectors back into Iraq" [56 percent]?

Notice: The question gave a reason to wait (for the United Nations to get inspectors back into Iraq) but no offsetting reason to act "fairly soon." A fair question would also have cited reasons *not* to wait. The *Times*'s pollsters were, I think, flagging to respondents their own attitudes; their questions clearly implied that one choice was right, the other wrong.

Encouraged by the *Times*'s antiwar campaign, allegedly confirmed by the poll, a parade of politicians and former Democratic presidents began to hammer at Bush's Iraq policy.

In September 2002, echoing the *Times*'s critique, former President Clinton warned that a U.S. attack on Iraq could prompt Saddam Hussein to use chemical or biological weapons against the United States. Calling for a focus on al Qaeda and Afghanistan instead of Iraq, Clinton said that, despite Saddam's weaponry, he "didn't kill 3,100 people on September 11." He said we should focus on the 9/11 terror network, saying "Osama bin Laden did, and as far as we know he's still alive."

The same week, former President Jimmy Carter criticized Bush's Iraq policy, as if rehearsing for the Nobel Peace Prize he was awarded later that year. Saying that Iraq posed no danger to the United States, he also stated that U.S. unilateral action "increasingly isolates the United States from the very nations needed to join in combating terrorism."

Carter's comments found their echo in a *New York Times* editorial of October 15, which read: "The war against terror requires Washington to build and lead a broad coalition, using diplomatic as well as military tools, and hold it together for many years to come. It is unclear how war with Iraq will affect this endeavor, but the events of the last few weeks [the Bali bombing in Indonesia] are a reminder that it is likely to make things harder rather than easier."

Former Vice President Al Gore joined the fray. Saying that he intended "to advance debate on a real important challenge that we face as a country," Gore accused the Bush administration of diverting national attention from the war on terrorism by his focus on Saddam Hussein. He complained that Bush had not done enough to seek international support for action against Iraq and had squandered the international goodwill America had after 9/11. "If you're going after Jesse James," he said cuttingly, "you ought to organize the posse first." (Why is Gore pithy only when he is wrong?) Continuing to echo the *Times*'s criticism, Gore accused the administration of leading "an attack on civil liberties" in its war on terror.

In retrospect, this speech probably made it impossible for Gore to run for president in 2004. The *New York Times* may well have helped to lead him astray.

On September 22–September 23, 2002, right in the middle of the national debate over war with Iraq, the pollsters struck again. This time, the survey, normally conducted as a CBS/*New York Times* poll, was sponsored only by CBS News. But the bias remained the same: the results were published under this headline: "No Rush to War." Once again, the pollsters mistook the message of their own poll. In the earlier September survey, respondents were asked if they thought that ". . . the Bush administration has clearly explained the United States position with regard to possibly attacking Iraq or haven't they done that yet?" At that time, only 27 percent felt the president had clearly explained his position, while 64 percent felt he had not.

Now, three weeks later, CBS repeated the same question. In the interim, Bush had presented his case forcefully in front of the United Nations and the world. This time, respondents felt Bush had clearly explained his position by 51–42, a complete reversal of the earlier data. But neither the *Times* nor CBS made any mention of the turnaround in public attitudes.

The public liked what they heard from the administration—but you'd never know it from the slanted account CBS gave. Poll respondents approved of Bush's handling of the war on terror by 71–22, up from 68 percent approval three weeks before. Instead of featuring this strong and rising support for military action against Hussein, CBS emphasized respondents' answers to a series of loaded questions to justify its headline that voters opposed a "rush to war."

For example, it attached great importance to the question "Is Congress asking too many questions about President Bush's policy toward Iraq, or isn't Congress asking enough questions?" Noting that respondents felt, by 44–22, that Congress wasn't asking enough questions, CBS noted that "there is broad support for getting Congress involved in the current debate about how to deal with Iraq. Twice as many Americans think members of Congress haven't asked enough questions about Bush's policy toward Iraq as think they've asked too many. Many Americans want Congress to take its time on this issue. . . ."

Since when don't Americans think that Congress should ask questions? The very phrasing of the poll question was designed to produce the answer it got. The theme of "take your time" came to predominate in the polling. But voters were never given a reason not to take time, so their support for delay was preordained by the question's wording.

But no matter how CBS and the *Times* massaged their data to produce the impression of waning public enthusiasm for Bush's plans on Iraq, they could not alter public opinion itself. So when the Democratic Party followed the *Times*'s data into a posture of opposition and questioning, it sealed its fate in the fall elections.

October 2002

Into the Stretch, the Times *Goes Back to the Economy*

As the midterm elections approached, the *New York Times* went further than it ever had to see in its own poll findings justification for its political agenda—to focus the elections on the economy rather than on the war on terror.

Iraq hadn't worked as a campaign issue for the *Times* and its Democratic allies. Despite the newspaper's attempts to raise doubts about the war and warn of its possible consequences, President Bush did indeed ask for congressional permission to go to war (as the *Times* had urged)—and got it, by 296–133 in the House and 77–23 in the Senate.

Bush also promised to ask the United Nations for a resolution authorizing inspections and, eventually, military action in Iraq. He got that, too, three days after the midterm elections. Bush had defanged the issue of American unilateralism, which the *Times* and the left had been ginning up for months.

And so it was back to the economy, as the newspaper scrambled to push the theme of its beloved Democrats.

The *Times* headline of October 7, 2002, could have served as a keynote for the Democrats' attempt to regain control of the House and preserve its majority in the Senate. Reporting the findings of its survey, the paper said, *on page one*: "Public Says Bush Needs to Pay Heed to Weak Economy; Many Fear Loss of Jobs: Poll Finds Lawmakers Focusing Too Much on Iraq and Too Little on Issues at Home."

Basing its conclusions on a heavily weighted poll conducted on

October 3–October 5, 2002, the *Times* announced its survey had found that ". . . the nation's economy is in its worst shape in nearly a decade and that President Bush and Congressional leaders are spending too much time talking about Iraq while neglecting problems at home."

But when you consider how the *Times* framed the question, it's hard to imagine what other result was possible: "Regardless of how you intend to vote in November, which would you like to hear the candidates talk more about, the possibility of war with Iraq or how to improve the U.S. economy?"

But those weren't the options in real life. It wasn't just the "possibility of war with Iraq," but the entire war on terror and the issue of national security, that had to be balanced against talking about the economy. When the question was asked another way, later in the poll— "Which of these should be the higher priority for the nation right now—the economy and jobs or terrorism and national security?"— voters opted by 50–35 for a higher priority for terrorism and national security.

The question also ignored the fact that many voters didn't care to hear candidates talk about Iraq because they'd already made up their minds that military action was necessary. After all, if they'd already decided about Iraq in Bush's favor, wouldn't they have been just as happy to start talking about the economy instead?

But the *Times* reported the results of the stacked question and buried the answers to the fair one.

In another loaded question, the *Times* asked whether voters would be influenced by "foreign policy issues" (25 percent chose this option) or "positions on the economy" (57 percent opted for this choice). What a choice! "Foreign policy or the economy"! If the question had been "terrorism or the economy" or "national security or the economy," the outcome would have been quite different.

But the *Times* took the loaded question and concluded that ". . . no matter what is happening in Washington, voters are more concerned with the economy and domestic issues than what is happening with Saddam Hussein."

So the entire thrust of the *Times* story was bogus. And yet the paper trumpeted its conclusions in an influential front-page story—despite having evidence in the poll that ran directly counter to the story.

From the beginning, the *Times* had been pushing the idea on Americans and their opinion leaders that the economy was getting worse. Yet somehow this concern never caught on with the American people.

In fact, in the October poll, respondents said, by 59–39, that it *wasn't* getting worse. Asked how the economy had changed recently, only 39 percent said it was "getting worse," 46 percent said it was "staying about the same," and 13 percent felt it was "getting better"— for a total of 59 percent who felt it wasn't deteriorating.

Even if the *Times* had been correct in encouraging politicians to debate the economy, it still wouldn't have helped the paper's Democratic friends. Asked in October which party would do the best job of making "sure the country is prosperous," voters split evenly, with 38 percent citing the Republicans and 41 percent the Democrats—scarcely a margin on which to base an entire campaign.

By contrast, when asked which party would do the best job of "dealing with terrorism," the Republicans held a distinct 47–28 edge in the *Times*'s October poll.

The *New York Times* weighted this survey more than any other that it had taken, adjusting the number of Republicans in its sample down by 3.2 percent while adjusting the number of Democrats up by 1.8%, for a total "adjustment" of 5.0 percent.

Meanwhile, the editorial page of the *Times* also found what it wanted in the October survey. The fact that 67 percent of Americans said they supported military action against Iraq in the poll didn't stop the editorial page from reporting, on October 8, that voters were "wary of war." Ignoring its own data, the editorial said that the "*New York Times*/CBS News poll published yesterday . . . support[s] the sense of many around the country that Mr. Bush still has work to do if he hopes to persuade Americans of the need to use military force to disarm Iraq." The editorial went on to recall that "not quite four decades ago, Lyndon Johnson learned to his and the nation's sorrow that taking a reluctant country to war can severely damage the body politic. President Bush must be mindful of that danger as he draws the United States ever closer to military conflict with Iraq."

Of course, the results of the midterm elections of 2002 were a dramatic win for the Republican Party. Voters did, indeed, value Bush's leadership, particularly on terrorism, and they either didn't care as much about the economy or did not lay the blame for its sluggishness at the president's door.

In the end, the ultimate rejection of the *Times*'s campaign was the Republican victory of 2002. Despite the *Times*'s focus on Enron and the other corporate scandals and its attempt to link them to Bush, the majority voted Republican. Seemingly ignoring the *Times*'s warnings about the imminent threat to their civil liberties posed by Bush's war on terror, the voters backed the president. Overcoming the *Times*'s concern that Bush might usher in a new era of unilateralism, leaving America detested abroad, voters decided to give the Republican congressional candidates a decisive victory.

And how did the *Times* react to all of this?

In its postelection poll on November 24, 2002, the paper proved ever faithful to its Democratic clients: it did all it could to throw cold water on the Republican victory. The *Times,* in its front-page headline, labeled the results as "Positive Ratings for the GOP, *If Not Its Policy*" (emphasis mine). Forced at last to agree that the Republicans had won the election, the paper turned to its own postelection polling—and found no evidence that voters agreed with the party for which they'd voted!

How to explain the fact that the Republican Party *gained* in both houses of Congress—a feat accomplished by the president's party only three times in the sixty-nine midterm elections since the Civil War? The paper conceded that the Republicans won because "Americans hold favorable views of the [Republican] party and President Bush," but claimed that ". . . they are less enthusiastic about some of the policies Republicans are promoting according to the latest *New York Times*/CBS News poll."

The *Times* assured its worried Democratic readers: "The [poll's] findings suggest limits to the mandate that some Republicans have claimed for Mr. Bush as a result of the Republican sweep of the November elections."

And why did Bush win? The *Times* had trouble figuring it out.

Was it the tax cuts? No; the *Times* reported that "Mr. Bush's enthusiasm for his $1.25 trillion tax cut plan is . . . not entirely shared by the public."

Oil drilling in Alaska? No, voters opposed it.

Was it a demand for faster judicial confirmations? Nope, the electorate liked the idea of Congress taking its time.

Did Americans want one-party control of the White House and Congress? No, by 41–36 they wanted divided control.

Would Republicans run the economy better? No, only 28 percent thought so. Will taxes go down? Only 14 percent said yes.

So what was it that brought about victory for the GOP?

Incredibly, in the entire *New York Times* story of November 26, 2002, analyzing the reasons for the Republican victory, there was not one single mention of the word "terrorism." The closest the paper came was to allude, ever so genteelly, to the fact that "just over half [of the respondents] said they were confident in Mr. Bush's ability to handle an international crisis."

Once again, the newspaper missed the point. To read the story on the newspaper's postelection poll, one would never have guessed that the Republicans won the election of 2002 because of terrorism and national security.

The truth, of course, is that voters backed Bush's candidates because they supported the relentless focus on terrorism the president had shown since 9/11. They feared that the Democrats would dilute this focus and reduce the priority given the war on terror (as the *Times* had, indeed, buried the issue in its postelection survey).

Almost everyone else in the United States saw Bush's victory as a ratification of his stance in the war on terror. But not the *New York Times*. The Democrats may have been gracious in defeat, but Raines sure wasn't.

Oh, brother!

The Gray Lady Lets Her Slip Show

I first became aware of bias at the *Times* while I was working for President Clinton in 1996. Responding to a phone call from the office of Joseph Lelyveld, Raines's predecessor as managing editor of the *New York Times*, I agreed to a meeting to "get to know one another."

When he arrived at my hotel suite in Washington, accompanied by the paper's White House correspondent, I admit to being a little awed. The *Times* had been my daily guidepost since childhood. I doubt that my father—an eminent New York lawyer—would have believed it if our city had been leveled by a nuclear blast, unless it was reported in the next day's *Times*. If the *Times* was a secular bible, my dad was an evangelical, hanging on a literal interpretation of its every word. Now here they were, the paper's editorial lights, descended from Mount Sinai to meet with me.

As Clinton's chief political adviser, I knew their visit had something to do with the White House. But I was surprised when they asked me to help them to get an exclusive interview with the president. "We've tried for months and come up empty," the editor pleaded. "Can you help get it done?"

I spoke of Clinton's sensitivity to criticism from the *Times* and how he had bristled particularly at Howell Raines, who ran the editorial page.

A worried frown clouded the editor's formerly sunny face. "You know," he assured me *sotto voce,* "we don't think that the public cares about what happened back in Arkansas."

Wow.

I wondered if I heard right. Did the managing editor of the *New York Times* just imply that they'd pull their punches over Whitewater, Paula Jones, the Rose Law Firm, Hillary's billing records, the Web Hubbell hush money, and the rest of the scandals that had emerged from Clinton's Arkansas Pandora's Box—all in return for an interview???

I certainly got that impression.

The next day I was in the White House residence, after our weekly strategy meeting, whispering in Clinton's ear about my conversation with the *Times.*

"They're B.S.ing you," the president said. (He didn't use the initials.)

"No," I protested. "I wasn't fishing for the concession, they just threw it out."

"Hummmf," he grunted, moving on to our next topic.

Somehow, the interview got granted.

Then my phone rang. It was the reporter who had sat with his editor in my hotel suite. I'd known him for some time, and he was calling to tell me that he would be conducting the interview. I congratulated him, and he invited me for a drink.

As I crossed through Lafayette Park to get from the White House to the Hay Adams Hotel to meet him, I wondered why he had wanted to talk before the interview.

After some light chatter over drinks, he began, casually, to tell me the questions he was going to ask. "I'll ask him what are his proudest achievements, what he's most ashamed of, why he thought he lost the Congress [in the 1994 elections], what he proposes to do about Bosnia. . . ."

A reporter briefing a White House aide on the questions he was preparing to ask the president: this was about as common as it is for Nebraska to brief Miami on their football signals before the game. I couldn't believe it.

Pushing my luck, I prompted him. "Why don't you ask about . . ."

"Good idea," my obedient reporter/friend said as he jotted down notes.

The briefing before the interview wasn't even hard. Sitting on the couch with the president in his wing chair on my right in the Oval Office, I fed the reporter's questions to Clinton, and we worked out answers.

"What if he asks about Whitewater?" Clinton wondered.

"He won't," I assured him. "He's told me exactly what he's going to ask."

Clinton couldn't believe his luck!

Knowing exactly what was coming, we came up with answers to hit the ball out of the park. And, on May 19, 1996, Clinton's smiling face adorned the cover of the *New York Times Magazine*, over the headline "Facets of Clinton." The story touted the president as "one of the biggest, most talented, articulate, intelligent, open, colorful characters ever to inhabit the Oval Office." The story went on to call him "breathtakingly bright" and even noted that he "exudes physical attraction."

Perennially unsatisfied, of course, Clinton dwelt at length on the few slight jabs the article directed at him. But any casual reader might have mistaken the story for a paid ad in the magazine section.

The suspicions I began developing in the Clinton years have been confirmed under Howell Raines's tenure: the new *New York Times* doesn't deserve the trust we place in its impartiality and fairness. Its compass no longer points true north.

What's the answer? Not to stop reading it but to stop believing everything it says. For years we've had the luxury of being able to trust one newspaper to tell us the full and unbiased truth; that era is over. Now we must realize that even our most highly pedigreed news comes with propaganda, freely and even ingeniously mixed together.

We must sharpen our sixth sense of when we are being manipulated by the news media. To be truly well informed, we must learn to take our *New York Times* with side helpings of a few other sources, just to offset its slant. You can't trust the editorials, and you can't trust the news, in the new *New York Times*.

TWO

AFTER IRAQ:

THE MEDIA CREDIBILITY GAP

The war in Vietnam spawned what journalists dubbed a "credibility gap," as first Lyndon Johnson and then Richard Nixon lied their way through the war. With each exaggerated body count of supposed enemy dead and every prediction that there was "light at the end of the tunnel" and the war would soon be won, the credibility of the president and the presidency dropped lower and lower.

The war in Iraq has also produced a credibility gap, but this time it's not official Washington that Americans disbelieve—it's the media itself that we can't trust to tell us the truth. We have learned not to believe our news organizations, because we finally have the ability to watch something we were never meant to see: actual events as they're really happening, side by side with the spin the news organizations put on them.

Never before has the American public had the opportunity to watch and read the reports of the establishment news media at the very hour that the events themselves were unfolding in front of them live on television. Their collective understanding of the dissonance between the two is breeding a distrust of the major news organs—the broadcast networks and the major newspapers—that will probably long outlast the war.

Those of us in professional politics take the distortions of the media news for granted, and have even learned to play it, through what has come to be called "spin." We know what's actually going on in Washington, in the White House, and before Congress. Each evening, as we see how these events are covered by the news anchors and read what's written about them in the morning's press, we can't help but notice the difference between the reality and the coverage.

Until now, the average American has rarely, if ever, had that same opportunity. Now that they have, their reaction to media news is likely never to be quite the same.

Each morning, we sat reading our copy of the *New York Times*, the *Washington Post,* or the *Los Angeles Times* and ruminated on their prophesies of doom and quagmire. Then we all looked up from our morning paper to see television correspondents actually *embedded with* our troops, reporting quick advances, one-sided firefights, melting opposition, and, finally, welcoming crowds.

Then the television would cut back to the anchors and military analysts far from the battlefield. There, with their pointers and maps, we heard all about how we had too few troops in Iraq, and how the war plan had misfired, and how Bush's failure to enlist Turkish cooperation was likely to prove disastrous.

For months before the war started, we had read articles in the establishment media about how house-to-house fighting in Baghdad would consume our troops like a meat grinder. We heard dire television predictions of poison gas, missile attacks on Israel, and burning oil wells. None of it happened.

Then, as the war unfolded, it became obvious that the media and news organizations would seize on each minor mishap, every slight delay, and any variation from the best-case scenario as evidence of major catastrophe ahead. Deeply against the war to begin with, they emphasized casualties from friendly fire, the accidental deaths of journalists, temporary supply shortages, and the unavoidable killing of civilians, while downplaying the real progress made each day by the brave men and women who were on the ground fighting.

Who can forget Peter Jennings's belittlement of the joyous mob that hauled down Saddam Hussein's forty-foot statue, in a scene reminiscent of the fall of the Berlin Wall, as the action of a "small crowd"?

The disjuncture between the reality and the reporting became obvious to anyone who had eyes and ears. The establishment news media had been given our trust, and it had abused and insulted that trust. As one result of this credibility gap, the era of the Big Three—CBS, NBC, and ABC—as the dominant forces in the news business is finally ending.

This trinity, which has controlled American news and politics for the past forty years, had already lost much of the power it possessed in its heyday. In the Vietnam years, Lyndon Johnson needed only three television sets in the Oval Office to follow the news as his voters were getting it from the TV networks. With one set tuned each to CBS, NBC, and ABC, he could learn all he had to about what America was watching, what we thought, and what we knew.

But now Americans are no longer content to wait until six-thirty or seven in the evening for their majesties Dan Rather, Tom Brokaw, and Peter Jennings to appear and dispense their highly edited and profoundly biased summaries of the day's developments. Now, en masse, viewers are switching to the cable news networks—CNN, MSNBC, and, above all, the Fox News Channel—for the raw information they want.

As the war unfolded, viewers voted with their remote controls to end this triopoly. CBS viewership dropped 15 percent from its prewar totals, ABC fell 6 percent, and NBC gained an anemic 3 percent—while the Fox News Channel audience rose 236 percent, and CNN and MSNBC's smaller audience numbers recorded similarly impressive gains.

On morning television, the cable show *Fox and Friends* actually drew 2.9 million viewers, more than CBS's 2.8 million on its *Early Show*—the first time (but not the last) that a cable news station has beaten a network news program in ratings.

Among younger viewers (eighteen to thirty-four), the future portends even more dramatic changes: *CBS News* fell 15 percent, while Fox viewership increased fivefold.

War often produces major changes in media and news reporting. The Civil War saw the birth of photography as a journalistic tool. In World War II, Edward R. Murrow brought radio into its own with his dramatic reports of the Nazi blitz on London. In Vietnam, television became pivotal, as images of bloodshed soured American backing for the war. And the Gulf War of the early 1990s saw the growth of CNN, as all-news cable television became essential.

But the changes that are likely to ensue from the Iraq war are going to be even broader and more substantial. This has been a rough war for tyrants and those who try to control the thoughts of their people—in Baghdad but also in the American media capitals, at the headquarters of NBC, CBS, and ABC, and at the major newspapers struggling creakily to compete with their twenty-four-hour cable rivals.

OFF WITH THEIR HEADS!

. . .

The establishment news media had always opposed the war in Iraq. Before the first bombs fell, it demanded U.N. approval for the operation; then, when the attack started without it, its political opposition morphed into military skepticism and dire predictions of disaster.

As the war began, the news media focused intently on the blunder that, it said, would doom the entire campaign: the refusal of Turkey to allow coalition troops to use its territory as a staging ground from which to attack northern Iraq. R. W. Apple Jr., writing in the *New York Times*, called the situation a "debacle." In London, the *Independent* warned hysterically that the battle "plan perished when Turkey refused to allow US ground troops to use its bases." Then, as our troops raced through the Iraqi desert, bypassing towns and cities as they rushed toward Baghdad, the media told us that the military had made what might prove a fatal mistake in opening up our supply lines to harassment by enemy guerrillas left behind in the dash to the enemy's capital.

No less a military authority than CBS's Lesley Stahl lectured Secretary of State and former chairman of the Joint Chiefs of Staff Colin Powell on the March 26 edition of *48 Hours* that the American "rear was exposed" and our supply lines in danger. Powell respectfully dismissed her claim as "nonsense," putting her concern into perspective: "Every general who ever worked for me is now on some network commenting on the daily battle and, frankly, battles come and wars come and they have ups and downs, they have a rhythm to it."

At the *New York Times*, R. W. Apple noted that "with every passing day, it is more evident that the allies made two gross military misjudgments in concluding that coalition forces could safely bypass Basra and Nasiriyah." Apple predicted that Saddam's guerrillas would bog down our forces as they advanced to Baghdad. "As Mao famously said," he noted, "the populace constitutes the water in which the guerrillas can swim like lethal fish. In city after city, they are swimming,"

And when Saddam's troops fought back, the prophesies of doom became ever louder. On March 25, the *New York Times* editorialized that "Iraq's best soldiers seemed in no mood to lay down their arms"— a claim the paper cited as "the latest evidence that some of the initial hopes—even assumptions—that Iraqi resistance would quickly crumble seemed not to be panning out."

The *Times* reported that there had been no "mass defections" of Iraqi troops and that "thousands of fedayeen militia fighters and civilian-garbed security forces remain in the south, ready to cause trouble behind the advancing lines." Saying that our troops were "bogged down in tough fighting," the paper warned darkly that these events "provided a hint of the difficulties that may lie ahead in Baghdad."

ABC's Ted Koppel reported on the March 25 edition of *Nightline:* "Forget the easy victories of the last twenty years. This war is more like the ones we knew before."

Those who saw and read these pessimistic forebodings naturally wondered if all had gone awry. Like stalwarts echoing the party line, the news media descended on the central culprit of all that had, it said, gone wrong: the military's battle plan.

The plan was flawed!

The cover story of *Newsweek* on April 7 was headlined "A Plan Under Attack." "Did we start the war with enough force?" the magazine asked, noting that "as the blame game begins, the fight in Iraq is about to get a lot bloodier."

The critics were everywhere, each repeating what the other had said, only more ominously, the tale growing with each telling. *Newsweek* warned that "some air-power advocates [are] grumbling that the initial phases of the shock-and-awe campaign went too easy on the Iraqi capital." Maureen Dowd pointed a finger at Defense Secretary Donald Rumsfeld in the *New York Times.* "Rummy was grumpy. TV generals and Pentagon reporters were poking at his war plan, wondering if he had enough troops and armor on the ground to take Baghdad and protect the rear of his advancing infantry." She explained that the "cocky theorists of the administration, and their neo-con[servative] gurus, are now faced with reality and history: the treacherous challenge, and the cost in lives and money, of bringing order out of chaos in Iraq. With sandstorms blackening their TV screens, with P.O.W.s and casualties tearing at their hearts, Americans are coming to grips with the triptych of bold transformation experiments that are now in play."

The root of the problem, she wrote, was that "when Tommy Franks and other generals fought Rummy last summer, telling him he could not invade Iraq without overwhelming force, the defense chief treated them like old Europe, acting as if they just didn't get it. He was going to send a smaller force on a lightning-quick race to Baghdad, relying on air strikes and psychological operations—leaflets to civilians and e-mail and calls to Iraqi generals—to encourage Iraqis to revolt against Saddam."

Time chimed in with the headline: "Best-laid Plans: The Iraqi Army Has Been Neither Shocked Nor Awed. What the Allies Missed and How They Missed It."

The *Boston Globe* laid out the particulars of the indictment of Rumsfeld's war plan: "The Pentagon's worst mistake may prove to have been abandoning the Colin Powell doctrine of overwhelming force. Instead the Pentagon chose to feed troops into the battle piecemeal, in the so-called 'rolling start' strategy . . . there is no covering up the fact that the allies now find themselves thin on the ground trying to protect their long supply lines while advancing on Baghdad, with most of the towns in their rear unsecured. How useful would the Fourth Armored Division have been if it could have been redeployed to Kuwait following the Turkish debacle but before the ground war began."

On television, former army general and Clinton drug czar Barry McCaffrey joined the Sky-Is-Falling Brigade. "At the end of the day," said McCaffrey, "the question arises: Why would you do this operation with inadequate power? Because you don't have time to get them there? But we did. Because you don't have the forces? But we did. Because you're trying to save money on a military operation that will be $200 billion before it's done? Or is it because you have such a strong ideological view, and you're so confident in your views that you disregard the vehement military advice from, particularly, army generals who you don't think are very bright." McCaffrey warned that the United States "could take a couple to 3,000 casualties" in Iraq because Rumsfeld's war plan was flawed.

But the war was actually going quite well. The prewar predictions fell by the wayside with each day of the conflict:

- The Iraqis did not use chemical or biological or radiological weapons against our troops. Our military's campaign to dissuade enemy generals from using such weapons, warning of war-crimes trials by e-mail and cell-phone calls, had worked. The horror stories never materialized.

- There were no missile strikes against Israeli cities. As special forces units seized western Iraqi bases from which the strikes could have been launched, Israelis breathed easier.

- The oil wells didn't go up in flames. Coalition commandos seized them before Iraqis could execute orders to blow them up.

- Terror attacks on "soft" American and British targets throughout the world never happened.

And yet, in a sea of such good news, the media managed to find islands of negativity from which to continue their reporting.

Even more carnivorous than its American journalistic counterparts, the British press was scathing in its attacks. The *Independent* wrote that

> the strain is even starting to show on Donald Rumsfeld. . . . And small wonder. For he is the man in the hot seat as, eight days into the Gulf War of 2003, a once cocksure America is forced to face the possibility that it may be months, not weeks, before a war sold as a virtual cakewalk may finally be over.
>
> Did Washington, seduced by the dream of a speedy and easy victory, put too few troops in the field? 'No' of course, answer the architects of the strategy. . . . [But] esteemed experts beg to differ. They point out that the 250,000 troops deployed in and around Iraq are only half the force massed for Gulf War One—which moreover was fought in flat, desert conditions ideal for US mechanized armor. And the ground combat force is only between 75,000 and 100,000.

But the Pentagon had a hedge against the doomsayers. It had taken the unprecedented step of inviting television news correspondents to "embed" with coalition forces, sitting right next to the troops as they sliced their way forward. Their reports, often delivered via grainy new videophone technology, held Americans entranced as we watched the war to free Iraq ourselves, right before our eyes. It was like watching Neil Armstrong walk on the moon.

USA Today called the embedding system "a bold and proactive acknowledgement that welcoming media coverage under controlled conditions will play better than stories produced by reporters who are kept far away from the action and forced to depend on military accounts."

Rumsfeld put it in perspective: "We need to tell the factual story— good or bad—before others seed the media with disinformation, as they most certainly will."

One wonders: Was Rumsfeld talking about al-Jazeera or the American news media?

The story the embedded reporters told had nothing of the pessimism of the armchair reporters back home. As John Keegan, the defense editor of Britain's *Telegraph*, observed, "The older generation, particularly those covering the war from comfortable television studios, has not covered itself with glory." Keegan, who has a chair in military history at Sandhurst, Britain's West Point, noted: "Deeply infected with antiwar feeling and left-wing antipathy to the use of force as means of doing good, it has once again sought to depict the achievements of the West's servicemen as a subject of disapproval.... The brave young American and British servicemen—and women—who have risked their lives to bring down Saddam have every reason to feel that there is something corrupt about their home-based media."

Some complained that the use of embedded journalists blurred "the line between serious news and tabloid journalism." Peter Andersen, a professor in the School of Communication at San Diego State University, said that most Americans believe that the overwhelming war coverage is excessive." The professor wrote: "Thank goodness two networks showed the Oscars and NCAA tournament. While we need a respite from battlefield violence, we are seduced and captivated by this ultimate reality TV show."

But polling by the Pew Research Center emphasized that "eight in ten of those questioned" said that the six hundred embedded reporters traveling with U.S. and British troops were "fair and objective." Faced with the real reporting of actual events from the front and the pseudo-analysis of retired generals and journalists back home, Americans wisely recognized the value of live, firsthand information over prejudiced third-hand commentary.

Desperate to give the war a bad name, the news media deployed the ultimate weapon: the comparison with the dread war in Vietnam, which scarred a generation of Americans and left a deep suspicion of all things military and most things official. In their eyes, Iraq was suddenly becoming a new Vietnam.

Savage in attacking a battle plan she called "incoherent," Maureen Dowd of the *Times* asked why we should be surprised at the difficulty of winning the war: "Why is all this a surprise again? I know our hawks avoided serving in Vietnam, but didn't they, like, read about it?" It was hard, she said, "not to have a few acid flashbacks to Vietnam at warp speed."

Noting that "the hawks want Iraq to be the un-Vietnam," she predicted that the effort was doomed by a flawed plan that, she said, "some retired generals say is three infantry divisions short." Donald Rumsfeld, she charged, "is so enamored of technology and air power that he overrode the risk of pitting 130,000-strong American ground forces—the vast majority of the front-line troops have never fired at a live enemy before—against 350,000 Iraqi fighters, who have kept their aim sharp on their own people." On April 1, the *Washington Post* joined in ganging up on Rumsfeld, comparing his plan with the mistakes of the Vietnam War defense secretary Robert S. McNamara.

Columnist H. D. S. Greenway, writing in the *Boston Globe,* was most explicit in elaborating the Vietnam metaphor. "When the North Vietnamese first encountered American firepower in the valley of Ia Drang nearly four decades ago, they developed a new tactic called 'grab them by the belt.' What they mean was: get close enough to the Americans so that some of their firepower will be neutralized in such close proximity. The Iraqis learned that lesson in 1991 when they put all their chips on defeating the Americans in the open desert. This time they mean to grab us by the belt by fighting in the cities." But then events again confounded the critics. After a brief spate of attacks on the supply convoys creeping north to supply the troops outside Baghdad, the coalition forces learned to deal with the threat, and the trucks passed unmolested. The American army subdued the tenacious opposition in Nasiriyah and proceeded toward its objective—the heart of Saddam's empire.

As coalition air power shifted from command and control targets to pummel Iraqi Republican Guard units, they melted away, took off their uniforms, and threw down their arms. The British military efficiently and humanely secured the city of Bazra in the rear of our advancing troops, ending the threat of attacks from behind as they advanced.

The media, undaunted by the success of the war effort, found a new cause on which to fixate: Our troops weren't getting unanimous standing ovations as they rolled through southern Iraq. The *New York Times* noted that "so far the people [of Iraq] greeting American troops have been much cooler than many had hoped."

And the network hosts were quick to jump on this desert bandwagon. As he toured Iraq's capital with Saddam's secret police listening to every word, Dan Rather had already claimed that Iraqi women "are all Saddam supporters" because nobody dared tell him otherwise. Now, on *Good Morning America* on March 26, ABC's Diane Sawyer wondered: "What happened to the flowers expected to be tossed the way of the Americans? Was it a terrible miscalculation?" *Newsweek* called Vice President Dick Cheney's ultimately accurate prediction that U.S. troops would be hailed as liberators "an arrogant blunder for the ages." And Nicholas Kristof warned, in the *New York Times*, that "Iraqis hate the U.S. government even more than they hate Saddam. So if President Bush thinks our invasion and occupation will go smoothly because Iraqis will welcome us, then he . . . is deluding himself."

As soon as Saddam had fallen and could no longer torture and kill those who uttered a word of support for the Americans, the Iraqi people's true attitudes became clear. The parades, flowers, and welcoming women holding up their babies to be kissed by American troops materialized, just as had been predicted. Deputy Secretary of Defense Paul D. Wolfowitz had been right when the *Washington Post* reported that he argued that Iraqis are "a people that is still distinctly terrorized into silence. The Iraqi people are still not free to speak for themselves. Until this regime is gone, until the fear of Saddam and the other kinds of terrorists are gone, they're not going to be able to speak."

But still the media found new worries on which to fixate. Bloody Baghdad lay ahead!

Again, Dan Rather had laid the groundwork on January 24, warning on CBS's *60 Minutes II* that Iraqis were claiming that coalition troops "will have to wage a perilous battle in the streets of Baghdad." Rather noted that "the civilians we spoke with [during his visit to Baghdad] said they will fight, and warned that Baghdad's narrow streets and dark alleys were "a perfect place for Saddam to ambush the invaders."

As soon as the war was under way, the media piled on the Baghdad street-fighting story. On March 25, the *New York Times* editorialized on the same theme: "The climactic battles of this campaign will be in and around Baghdad, where the regime remains firmly in control and the city's air defenses, though degraded, are still functional. Unless there is a major break in the will of Iraq to keep fighting, American forces may be faced with the street-by-street fighting they had hoped to avoid."

The *Washington Times* described how John McWethy of ABC's *World News Tonight* told viewers on April 4 that his "intelligence sources are saying that some of Saddam Hussein's toughest security forces are now apparently digging in, apparently willing to defend their city block by block." Turning to anchor Peter Jennings, he said solemnly, "This could be, Peter, a long war."

"As many people had anticipated," Jennings, ever the parrot of the conventional wisdom, answered.

Why was the media so obsessed with the chances of failure?

Brent Baker, of the conservative Media Research Center, says, "I think the news media love to see failure. . . . In the months leading up to the war, liberal opponents said it would be awful, and that the Iraqis wouldn't love us, and there would be blood in the streets. So when the actual war started, they actually believed their own fear-mongering. When anything went even a little bit wrong, they said, 'Aha. We were right.' But that kind of negative reporting couldn't last long because reality outran it."

Of course Baghdad fell in a twinkling, as the statue of Saddam Hussein fell along with his vicious regime. The house-to-house street fighting never materialized, nor did the long casualty lists the media had predicted. Where TV commentator Barry McCaffrey had warned of three thousand U.S. casualties, by the end of the war in April there were about a hundred Americans dead.

Was the media guilty of treason? In some cases, there's a strong argument that they were.

The most famous was the case of journalist Peter Arnett, CNN's standby in the 1991 Iraq War. This time, reporting for NBC News and

National Geographic, Arnett actually appeared on Iraqi television to denigrate the U.S. war effort. Arnett told Iraqi viewers, "It is clear that within the United States there is growing challenge to President Bush about the conduct of the war and also opposition to the war. So our reports about civilian casualties here, about the resistance of the Iraqi forces, are going back to the United States. It helps those who oppose the war." Arnett was summarily released from his NBC and *National Geographic* posts soon thereafter.

When the *Los Angeles Times* ran an article by John Daniszewski headlined "Every Day Gets Worse and Worse," the paper came in for a harsh and justified attack by Fox News's Bill O'Reilly. The story's headline came from a quote from an Iraqi woman. " 'Every day gets worse and worse,' Sahar, a twenty-three-year-old with a birdlike voice, said with a sigh on Wednesday. 'I can't imagine what will be next week.' "

Who is Sahar? O'Reilly revealed the truth that Daniszewski omitted: She was the person who "had been assigned by the [Iraqi] Information Ministry to guide, translate, and keep an eye on foreign journalists."

"Can you believe it?" O'Reilly fumed. "This *L.A. Times* reporter is quoting a woman who works for Saddam about conditions inside Baghdad, and her quote is a page-one headline! Unbelievable. Why don't they just put Saddam's latest press release on the front page *L.A. Times*!"

It's one thing when an individual reporter acts contrary to America's national interest, but it's quite another when an entire network does. As soon as the war ended and Saddam was removed from power, CNN's chief news executive Eason Jordan disclosed that his network had "withheld details of Saddam's brutality from its coverage. Alarming facts about secret police, abductions, beatings, dismemberment and assassinations under the Iraqi dictator were not reported to the public," Mr. Jordan wrote, "because doing so would have jeopardized Iraqis, particularly those on our Baghdad staff."

Among the things Jordan kept to himself was a vital piece of information that might have helped our country. Uday Hussein, Saddam's son, had told him in 1995 that he wanted to kill his two brothers-in-law for defecting to Jordan—and King Hussein for allowing them to defect to his country. Because CNN and Jordan chose to sit on the story, the brothers-in-law, crucial sources of information to the United States about Saddam's arms programs, were lured back to Iraq and executed.

Other journalists did not act against the national interest so overtly but often found themselves doing so as they bent over backward to avoid being "spun" by those supporting the war effort. "The process of trying to get it right is weighing heavily on all of us," said Steve Capus, executive producer of *NBC Nightly News*. "We want to get it right; we don't want to be spun. We don't want to be unduly influenced by the images flying through the air, as remarkable as they might be."

Paul Slavin, executive producer of ABC's *World News Tonight*, said the U.S. Central Command "is definitely not giving out enough detail to support its case. They're giving the assertion that everything is going well, everything is according to plan, and it's very hard to accept at face value."

The *New York Times* reported that "such concerns have led some in television to question whether all the access was ultimately in the best interests of war journalism. 'These real-time images of combat are indeed compelling,' Jeff Greenfield said on CNN yesterday. 'The hard question is, do they inform us—or unintentionally mislead us?'"

In other words, the television elite media was asking, Should we put aside our preconceptions and watch what is actually going on?

Nah.

Indeed, some in the establishment media felt that their coverage was too favorable to the war effort! An op-ed in the *New York Times* by Lucian K. Truscott IV, a graduate of West Point and a novelist and screenwriter, warned of manipulation of the media by the embedding of reporters in combat units:

> When Secretary of Defense Donald Rumsfeld first announced that reporters would be welcome in the trenches, members of the media were suspicious. After all, this was the same Pentagon that kept journalists far from the front lines during the Persian Gulf War. Yet from reporters inhaling the exhaust of infantry units to bleary-eyed New York anchors spellbound by squads of generals analyzing the data stream, the news media have marched practically in lock step with the military.
>
> Not since the halcyon days of Ronald Reagan has an

administration been so adept at managing information and manipulating images. In Iraq, the Bush administration has beaten the press at its own game. It has turned the media into a weapon of war, using the information it provides to harass and intimidate the Iraqi military leadership.

But the embedded correspondents told the true story of the war. Their reports, often given amid the heat of battle, plainly showed the success of American forces and the warmth of the gratitude of the Iraqi people. Ultimately it was those real-time reports, shining through amid the media naysayers' commentary, that destroyed the credibility of our major media news outlets.

The difference between their reports and the stories and analyses that ran in the next day's newspapers and on that night's television created the credibility gap from which the establishment media now suffers.

The American media establishment has blundered and blundered badly. In its attempt to tell us what to think, it has let us see the strings with which it manipulates us, and we don't like it one bit.

The era of media dominance of our political system has lasted forty years, ever since the news organizations brought down first Johnson and then Nixon. But it lasts no longer. It has lost the one thing it could not afford to lose: our trust and confidence in its impartiality.

The triopoly of the networks and the transcendent importance accorded a few newspapers and their favored columnists were always artificial. Americans are too diverse a group to be controlled by a handful of reporters echoing the same phrases, words, and ideas. But the establishment has been a long time in dying.

Now, it appears, it has overstayed its welcome; its reign is ending. In its place will arise the cacophony of divergent voices that truly constitutes free speech and a free press. The mindless parroting of "news analysis" from one organ of the journalistic establishment to the other will be replaced by a true pluralism of opinion, nurtured by our distrust of any one source and made possible by the rapid spread of news through the Internet. Instead of a carefully conducted symphony of one opinion, coordinated and conducted by the elites at the *New York*

Times and the TV networks, we will hear each voice and each opinion and be free to make our own decisions.

That's as it should be in a democracy.

This war has brought freedom to Iraq, but it's bringing some new kinds of freedom to America, too.

THREE

APRÈS MOI, LE DELUGE:
HOW CLINTON LEFT TICKING TERROR TIME BOMBS FOR BUSH TO DISCOVER

As King Louis XV lay dying, he ruminated about the state of the pre-revolutionary French kingdom his son would soon inherit. Everywhere he looked, he saw peril—the anger of the peasants, the arrogance of the nobility, the unfairness of the tax system. Sadly, he reflected that the young man who would become Louis XVI faced tough times. Old King Louis may not have known that his son and his daughter-in-law, Marie Antoinette, would lose their heads to the guillotine, but he must have had some sense of looming catastrophe. *"Après moi, le deluge,"* he said. After me, the disaster.

As Bill Clinton left office in January 2001, America was outraged by his final insult to the integrity of his office—the pardons he granted to the rich, corrupt, and undeserving. But only months later we learned of his final, horrific insult to us—that he had bequeathed to George W. Bush three ticking time bombs that would shortly explode: al Qaeda, Iraq, and North Korea.

President Calvin Coolidge's name was forever blackened after the apparently prosperous economy he left his successor, Herbert Hoover, imploded in a stock market crash seven months later. If history is just, President Bill Clinton's will likewise be blamed for leaving George W. Bush a nation unaware of, and unprotected from, the deadly peril that hit seven months later.

How much did he know? Everything he needed to. Al Qaeda was no unknown force to Bill Clinton: The terrorist group had struck the United States repeatedly on his watch, bombing the World Trade Center, the U.S.S. *Cole*, two U.S. military barracks in Saudi Arabia, and our embassies in Africa. Iraq had kicked out U.N. weapons inspectors, and it was diverting most of the $2 billion per year it was getting in oil money to buy and develop arms. And American intelligence had found that North Korea was secretly building nuclear weapons in vast, underground caverns, violating a commitment it made to Clinton in 1994.

Clinton knew where all three time bombs lay. His national security people even briefed Bush's incoming administration on the dangers of al Qaeda as they left office. But Clinton had done little to catch al Qaeda; he'd done nothing to rein in Iraq; and he had actively covered for North Korea as it violated its treaty commitments.

As he left the White House, he could well have said: *Après moi, le deluge.*

To understand how to deal with the enemies we now face, we must look hard at the Clinton administration's failures to face them down during his eight years at the helm. This inquiry is not an exercise in partisan recrimination. Nor is it merely an opening salvo in the historical debate about Clinton's role in fighting terror. Rather, it represents an urgent attempt to pin down how we got into trouble, to help us in getting out of it. We need to grasp the causes of our current predicaments before we can grapple with the solutions.

But one thing we do know: For years, Bill Clinton swept these three problems under the White House rug as they grew more dangerous and more immediate. And so, lest he reach too early for the role of Democratic spokesman, I call upon us all to look at his record on terrorism and join me in calling:

OFF WITH HIS HEAD!

Back in 1996, in one of my last days in the Clinton White House, the president and I discussed how history would view him.

"I think your place in history will rest on three big things," I said. The president grunted, a cue to proceed. "First, I think you have to make welfare reform work. I think you have to implement the balanced budget plans you've laid out. And finally, I think you have to break the back of international terrorism, by economic and military action against the terrorist states."

President Clinton was in a philosophical mood as we chatted by phone that Sunday morning, August 4, 1996. He had just signed the welfare reform bill; now, poised for a big reelection victory in the fall, he wanted to talk about presidents, history, and his own administration. We discussed each of the forty men who had held the office before him, dividing the eighteen we liked the best into three tiers. That left twenty-two out in the cold.

"Where do I fit in?" he asked.

"Right now, to be honest, I think you're borderline third tier," I said, choosing my words carefully. "It's too early to rank you yet, but you're right on the cusp of making third tier."

"I think that's about right," he replied, to my relief. Clinton never liked sycophancy. Unless you criticized him as harshly as he usually did himself, he didn't take you seriously. "What do you think I need to do to become first tier?" he asked.

"You can't be first tier"—I broke the bad news gently—"unless unanticipated historical forces put you there."

"Like a war," he agreed. "Okay, second tier?"

I replied by reciting my list—welfare reform, balancing the budget, and fighting terrorism. "You had hoped to do it [break terrorism] with the peace process, but [Israeli Prime Minister Shimon] Peres's defeat closed that door. Now you have to smash it militarily, and through sanctions."

Clinton's welfare reform legislation has proven more successful than even its most ardent supporters had dreamed. Reducing welfare rolls by more than half, it has simultaneously led to an almost one-third reduction in poverty. He not only balanced the budget but generated huge surpluses. Even after he left office, the United States was well on its way to paying off the national debt when the double whammy of the 9/11 attacks and the usual Bush family economic slump sent us back into red ink. But once the economy regains its footing and the terrorism crisis passes, the sound fiscal course on which Clinton helped to put the nation will likely continue where it left off in 2001.

But on the war on terror, Clinton was an utter and total failure. His record of inaction is bad enough, but his inability to grasp the dimensions of the issue, as I witnessed it in our conversations, was worse. In our time this may have become a trite phrase, but there's simply no other way to put it: He just didn't get it.

Clinton knew every statistic, argument, and nuance of the issues he had made his own—welfare reform, deficit reduction, student performance, Head Start availability, crime, export promotion, and so on. But on terrorism, during his first term—the period I witnessed firsthand—he knew little and cared less.

All our terrorist problems were born during the Clinton years.

It was during his eight years in office that al Qaeda began its campaign of bombing and destruction aimed at the United States. It was then that the terrorist group orchestrated its first attack on the World Trade Center; hatched a plan to destroy New York's bridges and tunnels and the U.N. building; conceived an effort to destroy eleven U.S. passenger jetliners; twice bombed U.S. bases in Saudi Arabia, killing nineteen Americans; bombed American embassies in Africa; and attacked the U.S.S. *Cole*. Bill Clinton and his advisers were alerted to the group's power and intentions by these attacks. But they did nothing to stop al Qaeda from building up its resources for the big blow on 9/11.

Iraq was a subjugated nation when Clinton took office. Recently defeated in the Gulf War, its military infrastructure was largely destroyed. But under Clinton's intermittent and easily distracted gaze, Saddam Hussein took the opportunity to rebuild his military, expel U.N. arms inspectors, and open a spigot to get the money he needed to rearm under the so-called "oil for food" program. Moreover, on Clinton's watch the Iraqi dictator was able to rekindle his efforts to build nuclear weapons and further develop other arms of mass destruction.

North Korea first signaled its interest in developing nuclear weapons in 1994, when the issue was whether or not it would permit inspectors from the International Atomic Energy Agency (IAEA) to monitor the disposition of spent nuclear fuel rods from its electric plant at Yongbyon, North Korea. The international crisis that followed reportedly led even President Clinton to contemplate a preemptive strike to destroy the fuel rods before they could be turned into fissionable material for nuclear bombs.

To defuse the crisis, former President Jimmy Carter traveled to Pyongyang to meet with North Korean leaders and see if a compromise could be reached. The agreement Clinton ultimately negotiated required North Korea to refrain from using the spent fuel rods to produce bomb-grade material and obliged them to accept IAEA inspection of the site.

In return, the United States, Japan, and South Korea agreed to join in financing nonnuclear power plants in North Korea and to ship fuel and food to that beleaguered nation.

But Clinton was so eager to declare victory that he failed to monitor the enforcement of the deal as he should have. Americans were shocked in October 2002 when North Korea admitted it hadn't kept its end of the bargain—and was manufacturing fissionable material at a secret underground location.

All three critical situations America faces today—al Qaeda, Iraq, and North Korea—were either incubated or exacerbated on Bill Clinton's watch.

As I first became aware of this situation, I believed Bill Clinton was guilty of negligence and oversight. As I read the evidence, however, the picture darkened significantly. Clinton's attitude probably started as neglect of global terrorism—a field alien to the Arkansas governor's experience and worldview. But as his administration evolved and entered its second term, its failure to deal with these three looming threats began to seem more and more conscious, even deliberate.

Sapped by the effort to resist impeachment, focused on burnishing his legacy through his phantom deal with North Korea, anxious to avoid the political risk of major military action on the ground against al Qaeda, and eager to avoid stirring up things in Iraq, Bill Clinton deliberately postponed dealing with this trio of threats so he could leave office under a seemingly sunny sky.

That Sunday in August 1996, as we chatted by phone, Clinton mused that he needed to win a war if he were ever to join the first tier of presidents. He seemed to lament that none was available. But there was: the war on terror. He just chose not to fight it.

Should he have seen the threat of terrorism coming? From the very start of his administration, he had a series of clues about how serious the terrorist threat would become—starting when a bomb exploded in the World Trade Center in New York City.

The World Trade Center Bombing: First Shot Across the Bow

President Bill Clinton's uneasy history with terrorism began thirty-six days after he swore to "preserve, protect, and defend the Constitution of the United States." On February 26, 1993, a terrorist bomb exploded in the B-2 parking garage under One World Trade Center. The blast was triggered by twelve hundred pounds of urea nitrate, found in fertilizer, and three tanks of compressed hydrogen. This attack, the first foreign terrorist bombing on U.S. soil in modern times, ripped a five-floor hole in the building, instantly killing six people and injuring a thousand others.

In later years, in subsequent attacks, we became accustomed to seeing President Clinton at the site of such tragedies, seemingly struggling to control his emotions, biting his lower lip and fighting back tears. But New Yorkers were spared that piece of theater as they tried to cope with the impact of the bombing. The president never visited the site of the attack; he did not attend any of the funerals of its victims. What he did was go about his public routine.

The day after the attack, Clinton used his regularly scheduled weekly national radio address to console New Yorkers. Promising "the full measure of federal law-enforcement resources" in apprehending those responsible for the bombing, Clinton vowed that "working together, we'll find out who was involved and why this happened. Americans should know we'll do everything in our power to keep them safe in their streets, their offices, and their homes."

Touring New Jersey four days after the blast, Clinton did not detour from his preplanned schedule in order to visit the World Trade Center site across the Hudson River. The *New York Times* reported that "although Clinton spent much of the day in northern New Jersey, he did not visit the site of Friday's bombing. Such a visit had apparently been discussed among White House aides, but officials in New York urged them to avoid it." An anonymous "senior administration official" told the *Boston Globe* that "Clinton had a full schedule in New Jersey, with no opening for a visit to the site in Manhattan."

Though he said he was "heartbroken" for the families of those killed in the blast, while in New Jersey Clinton assured citizens that "we've been very blessed in this country to have been free of the kind of terrorist activity that has gripped other countries. But I think it's important that we not overreact to it." He called on New Yorkers "to keep your courage up and go about your lives."

The *Globe* noted that "while security was noticeably tight during

Clinton's visit to New Brunswick and Piscataway, he did leave his limousine at one point to ride from the airport in Newark with some children going to a learning center." First things first.

Why didn't Clinton visit the site? The emphasis in his public statements and in the demeanor of New York officials in the aftermath of the attack was to avoid an "overreaction." Worried about public panic, and perhaps concerned that a presidential visit would get in the way of rescue and investigative efforts, New York officials told Clinton to stay away.

Okay, but what about afterward? President Bush let the smoke clear at Ground Zero for a few days after 9/11, but less than a week went by before he went and memorably addressed the rescue workers through a bullhorn, rallying them and reinvigorating America's sagging spirits. Bill Clinton, on the other hand, never visited the World Trade Center in the aftermath of the 1993 bombing.

He didn't go because he chose to treat the attack as an isolated criminal act, devoid of serious foreign policy or military implications. The fact that this was the first foreign terrorist attack on American soil seems to have set off no alarm bells at the young Clinton White House. The president treated it as a crime rather than as a foreign policy emergency. He defined terrorism as a law enforcement problem, not as a matter of national security. To Bill Clinton, it was not unlike any other homicide.

Commenting on the former president's approach to fighting terror, Bill Gertz, in his best-selling book *Breakdown*, underscores how the administration saw terrorism in the context of law enforcement: "The Administration's primary goal here [in response to terrorism], as always was to identify terrorists, capture them, and return them for prosecution in a court of law. It was a reactive strategy that did nothing to deter attacks."

Clinton wasn't the first to make this mistake. In his book *The Right Man*, Bush speechwriter David Frum notes, "In the thirty-three years before September 2001, close to one thousand Americans had been killed by Arab and Islamic terrorists. . . . Only once in all those thirty-three years did an American president interpret a terrorist atrocity as an act of war, demanding a proportionately warlike response: in April 1986 when Ronald Reagan ordered the bombing of Tripoli after Libyan agents detonated a bomb in a Berlin discotheque, killing two American servicemen."

Frum points out that "all the rest of the time, the United States chose to treat terrorism as a crime to be investigated by the police, or a clandestine threat to be dealt with by covert means, or an irritant to be negotiated by diplomats."

So dismissive was the White House of the 1993 attack that even after the same terrorist group—al Qaeda—had attacked the same target—the World Trade Center—a decade later, former White House adviser George Stephanopoulos minimized the 1993 assault, saying on December 30, 2001, that "looking back, it wasn't a successful bombing." He described the White House reaction at the time: "It wasn't the kind of thing where you walked into a staff meeting and people asked, what are we doing today in the war against terrorism?"

Obviously, they should have.

But there was no effort to mobilize the nation, to sound the alarm, to reequip the military and intelligence apparatus to cope with the new threat. The government did nothing. Indeed, the director of the CIA, R. James Woolsey, later said he had *not had a single private meeting* with President Clinton through all of 1993 and 1994. Incredible.

Had Clinton zeroed in on the terrorist threat, he would have certainly made time for the director of the CIA. To fail to see him one-on-one during a period of terrorism would be as ludicrous as to deny the head of the Office of Management and Budget a hearing during the government shutdown budget crisis of 1995.

Clinton's Twin Allergies: Foreign Policy in General, Military Action in Particular

President Clinton rarely had his mind on terrorism in the opening years of his White House tenure. In early 1993, upon taking office, he and the first lady were focused on how to balance the competing needs of stimulating the economy and reducing the budget deficit. What scant time there was to discuss military policy in those first days was devoted to how to keep President Clinton's campaign promise to allow homosexuals to serve in the military, a commitment he fulfilled in an executive order issued right after he took office. The military was sent into an uproar by the move; then-chairman of the Joint Chiefs of Staff General Colin Powell was reluctant to allow gays in the military without precautions to prevent a breakdown in discipline. Soon the debate was swirling throughout the nation.

Abroad, the administration was largely preoccupied by the need to extricate American forces from a poorly planned and ultimately futile mission in Somalia. As he was leaving office, President George H. W. Bush had ordered twenty-five thousand troops to that east African nation to check a massive famine, deliberately exacerbated by local war lords.

"His parting gift to us," First Lady Hillary Rodham Clinton called the involvement. She pushed the subject with her husband: "I keep telling him to pull them [the troops] out," she told me in a phone conversation in early February 1993, "but I have limited influence on foreign policy."

In fact, President Clinton himself felt he had similarly limited power over foreign affairs. Constitutionally empowered with virtual czarlike authority over international relations—in a way he never could be in domestic policy—Bill Clinton chose nevertheless to delegate most of his power to his two top advisers in the area: Secretary of State Warren Christopher and National Security Advisor Tony Lake. Concerned about his own lack of experience in foreign policy, Clinton began his first term by ruling over international affairs in name only, delegating any real power to the Christopher-Lake team.

Christopher is a lawyer; Lake is a liberal. The former was constantly ensnared in legalisms, the latter easily manipulated by the liberal foreign-policy establishment. Indeed, Lake had been given the office of National Security Advisor, in part, because the post required no Senate confirmation. When Clinton later nominated him to be director of the CIA, at the start of his second term, the Senate's investigation of Lake's liberal record eventually persuaded Lake to withdraw his nomination.

In any case, neither Lake nor Christopher was president of the United States. While they could exercise strong influence, neither could commit the American people to the kind of mobilization that would have been needed for an all-out war on terror. Only the president could do that, and he was extremely reluctant to act, given his limited grasp of the subject.

When I signed up as his adviser, two years into Clinton's first term, I teased the president about his reliance on the Christopher-Lake regime in foreign affairs. "I think I'm beginning to see how Lake runs foreign policy around here," I said in March 1995. "There's a regency," I observed, referring to the way European monarchies appointed adult ministers to guide underage kings. "You're too young now to run your own foreign policy, so Lake and Christopher have to do it. But when you turn twenty-one they'll let you take it over."

The president stiffened slightly at my characterization. But he said only, "I never get other options; I never get other information."

Clinton saw foreign affairs as a subset of economic policy, rejecting the cold war view that it was related to global diplomatic and military manifestations of power. Deconstructing von Clausewitz's famous dictum that "war is diplomacy by other means," Clinton saw diplomacy as economics by other means.

As governor of Arkansas, Clinton was accustomed to dealing with issues of world trade and he felt at home stimulating the domestic American economy by manipulating global commerce. But he had no experience in foreign affairs, and political fallout from his Vietnam draft-dodging experience had left him with an allergy to military action.

The draft issue had first surfaced early in 1992, as Clinton sought to win the New Hampshire primary. He was hit simultaneously by two scandals, a combination that threatened to deck his candidacy in the early rounds: his adultery with Jennifer Flowers and his avoidance of military service. Together, the issues swiftly erased his early lead in the nation's first primary.

As the New Hampshire campaign approached its apogee, Clinton called me in a panic while I was on vacation in France. With his typical charm, he apologized for calling me at seven A.M., Paris time: "It's one o'clock here," Clinton explained. "I stayed up as late as I could so I wouldn't wake you up too early."

Then he got to the point. "I'm getting killed by the draft thing."

"And Flowers," I added.

"No," he interjected. "Our polls show Flowers isn't really hurting, but the draft is killing me, killing me," he repeated. "How should I answer it?"

He proceeded to give me a detailed and tedious account of how he had, he said, used only legal means to avoid service, stopping well short of the "string pulling" of which his opponents were accusing him.

"I don't think you can win on the draft," I answered. "I wouldn't try. Anything you do to talk about it just makes it a bigger issue. You took the lead in New Hampshire because you ran substantive commercials with real programs and new ideas about welfare, the economy, and other topics. Just go back to that. Put out a positive, exciting message, and I think you can come in second," I said.

"You think that'll work?" he asked doubtfully.

I wondered myself, so I took another tack: "Look," I said, "you're getting hit with two charges—the draft and adultery. The draft is hurting you, but the Flowers stuff isn't. Answer the Flowers stuff and use it to drown out the draft issue. It's sexier anyway—and that way everyone will pay attention to the disease that's not fatal and ignore the one that could be."

He did just that, finishing second in the primary and promptly declaring himself the "Comeback Kid."

But the scar of his near defeat over the draft lingered, a silent inhibition that consistently held him back from aggressive military commitment. No matter how often his soldiers and generals snapped to attention as he passed, giving him crisp salutes, the reality remained— having ducked the draft, Clinton never felt comfortable with the prospect of sending young men and women to face death when he had refused to risk it himself.

The Emerging Terrorist Threat, 1993–1996

It is never fair to assess a president's conduct entirely in the light of 20/20 hindsight. The question for future historians is: How well did the chief executive act, given the information available to him at the time? So, in assessing Clinton's performance in preparing us for the war on terror, we must ask: What did the president know and when did he know it?

The story of the investigation that followed the 1993 World Trade Center attack is instructive. At first, the bombing seemed the work of a Laurel and Hardy band of incompetent terrorists. Having sought to topple the Twin Towers, they had succeeded only in making a big hole in its lower floors. Investigators quickly determined that the bomb had been planted in a Ford 350 Econoline van rented across the Hudson River in Jersey City. When detectives staked out the location, they had trouble concealing their shock when, two days after the bombing, Mohammed A. Salameh showed up to collect his $400 deposit on the van. (Terrorists on a budget!) Soon the other conspirators were identified and arrested, with the exception of mastermind Ramzi Ahmed Yousef, who had fled to Pakistan.

Yousef had entered the United States six months before the bombing, disguising his evil mission by pretending to seek political asylum. According to Steven Emerson's meticulously researched *American Jihad*, Yousef "soon emerged as an international link between the assailants and a network of supporters."

As Emerson recounts, Yousef had arrived in the United States "in the company of Ahmad M. Ajaj, a pizza deliveryman in Houston. Ajaj was detained at the airport for carrying three fake passports and other false identification." Ajaj was also found to be carrying "a letter of introduction recommending him for training in guerrilla warfare in Pakistan or Afghanistan." How, in the name of God, was he passed through to enter the United States?

Nevertheless, he was. And it is clear that, in the weeks after the bombing, this and other clues alerted the administration to the fact that the World Trade Center bombers were no isolated fanatics—but rather a group of conspirators with close ties to foreign terrorist groups. That alone should have put the White House on alert. As Richard Bernstein, who covered the bombers' trial for the *New York Times,* asked: "Who wrote Mr. Ajaj's letter of introduction? Why would he have to travel to the Middle East to obtain it? Whom did he see in the United Arab Emirates and Saudi Arabia? Where did Mr. Salameh, who was certainly not a wealthy man, get the $8,400 that he deposited into a bank account he opened jointly with Mr. Ayyad (another conspirator) and that prosecutors say was the bankroll for the operation? Did it come from abroad?"

An alert White House would have been all over these questions, weighing their implications for America and its future, and acting accordingly. The failure to heed these and other warning signs, and to mobilize fully our nation's resources to protect us against further acts of terrorism, are reason enough for a stinging indictment of the Clinton administration.

Clinton should also have known that the New York bureau of the FBI had been investigating Sheikh Omar Abdel Rahman, a blind Egyptian cleric who came to this country after facing charges of attempting to assassinate Egyptian president Hosni Mubarak (a charge of which he was acquitted). According to Emerson, the FBI was able to infiltrate an informer, Emad Salem, into the Jersey City group that surrounded Rahman. "Salem carried hidden microphones and helped the FBI in planting a small video camera, recording the group as it made plans for a Day of Terror."

Emerson relates how these terrorists planned nothing less than "simultaneous strikes at the United Nations headquarters, the Lincoln and Holland Tunnels, the George Washington Bridge, and the federal office building" in lower Manhattan. These bombs would have damaged or destroyed the only links that connect midtown and downtown New York with New Jersey—tunnels and bridges used by millions of people each day.

In June 1993, when the FBI arrested Sheikh Rahman and nine of his followers, President Clinton must have been told that the terrorist groups in and around New York City were actively plotting massive destruction of high-profile targets. The World Trade Center had already been bombed, the United Nations and bridges and tunnels had been targeted. What else did the president need to grasp the gravity of the situation? Yet he never ordered any major shakeup of the antiterror apparatus. No extra tools were given to the FBI. No massive mobilization was declared. The government simply shrugged its shoulders; *the bank robbers had been caught, after all; why make a fuss?*

The most important link in the chain of evidence that should have alerted Clinton to the growing threat came in January 1995, when Yousef himself was finally arrested in Pakistan, two years after orchestrating the World Trade Center bombing. Under interrogation, Emerson writes, the terrorist leader said he had "hoped [WTC] Tower One would fall sideways into Tower Two" as a result of the bombing, "knocking over both and killing 250,000 people."

More important, an examination of Yousef's laptop computer revealed that he had "also participated in a plan to blow up eleven American jetliners within 48 hours—a disaster that was only barely avoided by chance."

One would have imagined that, at the very least, the president would have responded to the evidence of such a plan with a major air-safety initiative. Even if he wanted to avoid alarming the traveling public and jeopardizing airline revenues, one would think he would still have moved vigorously to tighten security, concealing the reason for his actions if necessary.

Instead, there was nothing. No action, no proposals, no initiatives, no direction. It was as if nothing out of the ordinary had been unearthed by the FBI.

Why not?

Beyond Clinton's reluctance to engage the military, another factor was at play here: Bill Clinton was a one-thing-at-a-time president. In his White House, there was no back burner. Either an issue was in the fore-front, occupying his undivided attention, or you couldn't expect to find it on his radar screen at all. Indeed, so far was the issue off the presidential priority list that the index to Bob Woodward's chronicle of the early Clinton presidency, *The Agenda,* contains not a single reference to terrorism or the World Trade Center.

Lurching from one issue to the next, Clinton devoted his attention in these early years to learning the ways of Washington. At first his attention was riveted on his futile attempt to pass an economic stimulus package to "jump-start" the economy. Keeping to his campaign promise to focus "like a laser beam" on the economy, Clinton then turned to his economic program—a combination of tax increases and spending cuts—which did much to reduce the budget deficit and gave Federal Reserve chief Alan Greenspan the room he needed to drive down interest rates.

Clinton spent 1994 preoccupied with his anticrime legislation, a mix of longer sentences, more prisons, an expanded death penalty, and gun controls, which passed narrowly and had a significant impact in reducing crime over the ensuing decade. What time he had left was devoted to the fortunes of his wife's health-care reform proposals, destined to die an agonizing defeat in Congress.

As 1994 drew to a close, Clinton traveled to the Middle East to help cement a diplomatic accord between Jordan and Israel, advancing the legacy of the Camp David accords of President Jimmy Carter.

Clobbered in the 1994 midterm elections, Clinton scrambled to respond to the Republican program of budget and tax cuts and to find his footing amid a Congress suddenly in the hands of his enemies.

I worked in the Clinton White House from late 1994 through late 1996. From November 1994 through February of 1995, I held seven private meetings, of two to three hours each, with President Clinton. In January, we worked together, in private, for six hours, drafting his State of the Union speech to Congress. In the course of these meetings, we spoke of every major issue he faced in our attempt to cope with the challenges posed by the newly elected Republican Congress. We delved extensively into policy initiatives about crime, law enforcement, and gun control, and we talked at length about how to strengthen his hand in dealing with foreign-policy issues.

Yet, in all those discussions, Clinton never mentioned a single word about the terrorist threat that was gathering around America. He did not allude to the World Trade Center bombings or to any of the evidence of further terror plans that had emerged since. The subject just wasn't on his mind, despite massive evidence flowing in from America's law-enforcement agencies.

Simply put, Clinton had no time for terrorism. Notoriously unable to delegate responsibility, compulsive in controlling the actions of his subordinates, Clinton was too busy juggling other issues to address the threat terrorism posed during these opening years of his presidency.

That is, until the end of 1994, when he was forced to face the first major foreign-policy crisis of his presidency—the imminent acquisition of nuclear weapons by the rogue regime in North Korea.

North Korea: The Feel-Good Deal That Left Our Security Dangling

On October 5, 2002, a bombshell burst: North Korea acknowledged, as the *Washington Post* reported, that it "has been secretly developing nuclear weapons for years in violation of international agreements." One official in George W. Bush's administration called it a "jaw dropping" revelation. The North Koreans were unapologetic; indeed, they were "assertive, aggressive about it."

As the *Post* reported, American assistant secretary of state James A. Kelly had presented the North Koreans with detailed evidence "of a covert nuclear weapons program" during a visit to the isolated state on

October 3–October 5, 2002. After denouncing the allegations as "fabrications," the North Koreans "met through the night before deciding to reveal that the project had been under way for several years."

The revelation was doubly shocking in light of a widely hailed agreement the Clinton administration had signed with North Korea, in 1994, which was to have banned the development of atomic weapons or the diversion of fuel from North Korea's nuclear reactors. North Korea had agreed to suspend operations at its Yongbyon nuclear plant and to seal, under international inspection, used fuel rods that could have been reprocessed into bomb-grade material. The *Post* reported that the United States, Japan, and South Korea, in return, "agreed to arrange construction of two light-water nuclear power plants (whose fuel is less likely to be diverted to pernicious use) in North Korea and to provide 500,000 metric tons of heavy fuel oil annually to generate electricity until the plants are built." The $4 billion cost of the reactors, and the bill for providing the fuel, was to be shared by the three allies.

The United States had kept its end of the bargain. But the North Koreans, apparently, had gone behind our backs and begun to build bombs using uranium it had highly enriched at a secret underground complex. In December 2001, The CIA's National Intelligence Estimate, according to the *Washington Post*, reported that the rogue state had likely had one or two nuclear bombs as early as the mid-1990s. Right under Clinton's nose.

It was a breach that called to mind the Nazi and Japanese treaty violations of the years before World War II. Pyongyang's actions were high handed, arrogant, sneaky, and duplicitous.

Clinton had been duped. Big-time.

The crisis with North Korea that led to the 1994 agreement in the first place began at the start of Clinton's term, when North Korea, as rogue as a nation can get, diverted nuclear fuel from its Yongbyon reactor in 1989 and was planning future diversions to build up enough material to produce five atomic bombs.

Clinton's foreign-policy team gathered itself for its first major test since its early confrontation with Haiti's dictatorship. Their efforts thus far to forge a coherent approach to international issues were widely regarded a joke.

The *Washington Post* reported that the challenge from North Korea "is the first time that the administration has tried to forge a coalition for a major strategic purpose. . . . The success or failure of Clinton's effort will be a gauge of administration diplomatic skills, which critics have found wanting in negotiations with allies over Bosnia, with military rulers in Haiti over the return of democracy, with Japan over trade and with China over human rights." A Clinton administration official added that Korea "is the primary test for the administration of acting through the Security Council."

Clinton began by threatening to impose economic sanctions to get North Korea to behave. He hoped that China would privately cooperate, to pull the noose even tighter. The sanctions would bar Koreans living abroad—primarily in Japan—from sending money to their relatives in the North, a key source of foreign exchange for the isolated government; ban arms sales to North Korea; and end economic aid from the United Nations. If Pyongyang did not give in, then full economic sanctions would be imposed. South Korea enthusiastically agreed to join the sanctions, and after some hesitation Japanese officials said they'd go along. A deal seemed in the making. For once, Clinton was showing some backbone.

News reports after Clinton left office speculated that the president may have been considering a preemptive bombing of the Yongbyon nuclear plant where North Korea was storing its spent fuel rods to prevent reprocessing. One wonders if these reports are accurate or just posturing by former administration officials anxious to justify their actions while in office. At the time, Clinton's foreign-policy team downplayed the idea that a military option was under consideration. The *Washington Post* quoted "senior administration officials and independent analysts" as saying that the United States was "unlikely to initiate military action in Korea," saying that a conflict was "too risky." Among their concerns was that a preemptive strike might bring on a "radioactive explosion."

Clinton may have secured the agreement of Japan and South Korea to economic sanctions, but he may have been surprised to find he also needed former President Jimmy Carter's approval. Or so Carter himself seemed to believe: As if he'd never left office, the ex-president stepped in, uninvited save by North Korea, and cooled whatever resolution Washington might once have had.

Angering Clinton, Carter decided unilaterally to travel to North Korea to seek a way out of the crisis. Clinton had used Carter's services as a mediator earlier in his administration, asking the former president to smooth the way for the departure of the dictators who ruled Haiti and the restoration of democratic rule. Backed by a large naval and amphibious force waiting offshore to attack, the former president was able to orchestrate a bloodless transition on that troubled island.

But this time Carter's trip was forced on Clinton, presented to him as a fait accompli. When Carter made it clear that he was going to North Korea, Clinton had no choice but to bless the mission. "Frankly, he was going to go anyway and . . . we didn't want this to be some dispute," a senior administration official told the *Washington Post.*

Of course, there was one problem: Carter opposed administration policy. He was against sanctioning North Korea. Rather than traveling as an envoy of administration policy, in truth, he was looking to block it. But Carter's views had resonance within the administration's dovish foreign-policy team. As the *Post* reported, there was widespread fear among Clinton's top advisers that sanctions could provoke North Korea to oust U.N. inspectors and lead to a go-it-alone pursuit of nuclear weapons that might, they feared, "lead to possible war."

On June 16, 1994, Clinton and his top foreign-policy aides huddled in what the *Post* described melodramatically as "a grim council of war—discussing sending new planes, ships, and troops to South Korea for a possible horrible conflict" when the phone rang. It was Carter, calling from North Korea to tell him that he would "shortly appear on CNN to convey what the former president considered a dramatic breakthrough in the . . . dispute" with North Korea.

The scene must have been something to watch. The *Post* describes how "Vice President Gore, Secretary of Defense William J. Perry, Secretary of State Warren Christopher and the others filed into a cramped office [adjoining the Oval Office], equipped with a television set, to watch Carter. They were flabbergasted when the former president described [North Korean dictator] Kim's promises as a 'very positive step' and urged the administration to withdraw a two-day-old proposal for . . . sanctions against North Korea. 'It looked as if we were contracting out our foreign policy, like we were bystanders . . . and had totally lost control of it,' a White House official later recalled."

Gore urged everyone to calm down, put aside their anger, and coolly analyze what was going on. But Carter's move had halted whatever momentum there was for sanctions, and tough action was put on hold while the administration played out the Carter initiative, forced to do so, in part, by public hopes raised by the former president.

In reality, all Carter had gotten out of Pyongyang was what the *Washington Post* called "a small concession"—a "limited and somewhat vague pledge to Carter . . . that North Korea would leave international inspectors in place at North Korea's Yongbyon nuclear complex, and freeze its accumulation of plutonium if Washington entered high-level talks," an offer it had made before.

But Clinton and his people seized on the gesture and followed the path of negotiation, its former determination to impose sanctions ebbing precipitously. The North agreed to freeze its nuclear program and, as the *New York Times* reported, "to allow inspectors to monitor where some spent reactor fuel rods were being stored. In return the United States agreed to resume negotiations with the North Koreas without other conditions."

In reality, North Korea had conceded nothing. The spent fuel rods were so radioactive that, as the *Washington Post* reported, they "cannot be reprocessed anyway for awhile while [they are] cooled in a storage pond." Republicans in Congress pointed out that "if North Korea wants to, it can find a reason to withdraw from the talks in several months and have enough plutonium . . . for four to five nuclear weapons. . . . Such an outcome would make Carter and the Clinton Administration look like dupes."

And dupes they were, as they abandoned their plans for coercion and grasped at the hope of negotiations.

In the meantime, Bill Clinton's presidency was entering a period of crisis. The disorganization and left-leaning policies of 1993–1994 were alienating even the Democrats who controlled both houses of Congress. In August 1994, Hillary Clinton's vaunted health-care reforms went down to crashing defeat in the Senate. With congressional elections looming in November, Clinton needed to pull a rabbit out of his hat if he hoped to hold on to control of Congress.

North Korea was his rabbit.

In late October 1994, just weeks before the election, the United

States and North Korea struck a deal: Kim Jong Il, who had taken over as dictator of the North after his father Kim Il Sung died in July, agreed to "internationally monitored containment and eventual rollback" of its nuclear capability, as the *Washington Post* reported. In return, the United States, Japan, and South Korea agreed to provide food and fuel for North Korea and to fund its two light-water nuclear power plants.

Hailing it as a "gigantic political breakthrough," the *Post* breathlessly announced that the agreement "could end the specter of a rogue state's going nuclear."

From the start, however, there were reasons to doubt whether North Korea would keep its promise. Even as it celebrated the deal, the *Post* reported that "North Korea's record of treachery, its maintenance of a regime conducive to treachery and its leadership uncertainties compel great wariness." The Clinton administration was consequently careful to demand that the North Koreans "freeze and dismantle the graphite reactors [and] comply with the nuclear abstinence demanded under the Nonproliferation Treaty" before the United States had to deliver on its end of the treaty.

Republicans criticized the deal as a bribe to get North Korea to do what it had already committed to do as a signatory of the Non-Proliferation Treaty. Congress grew increasingly restive about voting the funds necessary for the implementation of the agreement.

Such concerns lay more or less dormant for four years, while Clinton basked in the apparent triumph of his diplomatic accord with North Korea. But suddenly, in August 1998, North Korea fired a multistate ballistic rocket over Japan without warning. While North Korea "subsequently insisted that it was only trying to send a music satellite aloft to celebrate two imminent joyous events—the regime's 50th anniversary and the formal accession to supreme power of Kim Jong Il," as the *National Review* reported, the missile spoke volumes about the nation's warlike intentions. Kim's regime, already in possession of the world's third-largest arsenal of chemical weapons, had now developed the ballistic missiles it needed to deliver them.

After he left office, President Clinton feigned surprise that North Korea was cheating and developing nuclear weapons despite its commitments in 1994. He told interviewer Larry King, on February 6, 2003, that "it turns out they [North Korea] had this smaller laboratory program to develop a nuclear bomb with enriched uranium." He might not have said so explicitly, but Clinton's implication was clear: The development was news to him.

Don't be fooled: The revelation of North Korea's perfidy may have been a surprise to the world, but it was no surprise to Bill Clinton. As early as 1998, the *Washington Post* reported, U.S. intelligence had warned that the rogue nation was developing bombs in secret underground locations. But Clinton did nothing; indeed, he assured Congress that North Korea was in compliance with the 1994 agreement so that it wouldn't cut off the purse strings that funded the U.S. end of the deal.

In mid-August 1998, the *New York Times* reported that the U.S. intelligence community had detected "a huge secret underground complex in North Korea" that might be "the centerpiece of an effort to revive the country's . . . nuclear weapons program." On August 18, the *Washington Post* reported that "U.S. intelligence analysts believe about 15,000 North Koreans are at work on a vast, secret underground nuclear facility, a development administration officials say may represent a decision by North Korea to abandon a four-year-old agreement to freeze its nuclear weapons program.

"Administration officials who have been briefed on the intelligence data, which includes imagery collected by spy satellites, describe a large-scale tunneling and digging operation in a mountainside about 25 miles northeast of Yongbyon, a former nuclear research center where North Korea is said to have produced enough plutonium for two nuclear weapons." Is it possible that the intelligence agencies of the U.S. government could find such a massive and crucial development and fail to report it to the president of the United States? Not in this world.

Indeed, the United States asked for a look at the underground caverns. North Korea first blustered and threatened and then asked for a cash payment of $300 million if the inspection failed to uncover suspicious work on a bomb. Later, the North Koreans said they wanted the $300 million up front for a one-time-only peek. Obviously the president would have had to have known about this exchange; indeed, he would have had to authorize the communication in the first place.

But, despite the evidence of massive cheating, Clinton did nothing and told Congress nothing was amiss. The *Washington Post* reported that "administration officials have told Congress that North Korea has not yet technically violated the [1994] agreement, despite its development of the [underground] caverns, because the Yongbyon facilities [identified in the treaty] have not been reactivated."

But Congress began to act. In late 1998, the Senate voted by 80–11 "to condition funding on a presidential certification that North Korea has halted all nuclear activities and has curtailed missile sales to nations classified by the State Department as supporters of terrorism." But Clinton continued to wink at North Korean aggressive moves, hoping against hope that the 1994 deal would remain in place.

Ever the master of semantics, the president, who had previously denied that what he did with Monica Lewinsky constituted "sexual relations," now maintained that the treaty with North Korea wasn't really a treaty at all.

The *National Review* reported that "American negotiators who hammered . . . out [the 1994 deal with North Korea] have repeatedly emphasized that it is not an 'agreement'—that it does not bind any party to specific actions or hold parties in noncompliance if given objectives are not met. 'Failure' of the 'Agreed Framework, consequently,' the officials maintained 'is very much in the mind of the beholder.'"

There is some evidence that North Korea diverted the fuel rods that it probably used to make atomic bombs even before Clinton took office. According to the congressional research service, North Korea shut down its nuclear reactor for seventy days in 1989, which "gave it the opportunity to remove nuclear fuel rods, from which plutonium is reprocessed."

Even if the crime took place on the first President Bush's watch, Clinton failed to address it in the 1994 Framework Agreement with North Korea, and he continued to send Pyongyang fuel and food even though he knew the regime may have already illegally developed nuclear weapons.

Clinton had wiped North Korea off his radar screen, never to return during his term. And there matters lay until Bush took office and discovered that North Korea had been industriously building nuclear weapons all along and likely had one or two in its quiver already.

By his willful blindness to North Korea's conduct and his wishful thinking that the regime would abide by the deal he had made with it in 1994, Bill Clinton had opened the door to one of the most serious threats to our national security since the end of the cold war.

When prompt action could have headed off North Korean noncompliance, Bill Clinton willfully and deliberately did nothing, allowing the North to build its bombs in its underground caverns.

And, as in so many other situations, he left the problem to George W. Bush.

The Crackdown That Didn't Happen—Clinton Refuses to Act to Deport Illegal Immigrants

"Make states issue driver's licenses [to immigrants] which expire when [their] visas do," I suggested to President Clinton on March 16, 1995, during a strategy meeting in the White House's East Wing. Noting that half of the nation's illegal aliens had evaded the system by overstaying their visas, I proposed a system providing for "automatic referral from motor-vehicles agencies to the INS" for deportation when routine traffic stops revealed drivers without licenses who were here illegally.

Raising these two issues—immigration and terrorism—with the president for the first time, I commented that, after all, it is through motor-vehicle law enforcement that most people come into contact with the police. If we could use that interface to catch illegal aliens, we could add mightily to the deportation lists. By interfacing the Immigration and Naturalization Service (INS) and motor-vehicle computers, we could determine, immediately, if an unlicensed driver was just a minor scofflaw or in active violation of immigration laws.

The INS had no organized way of identifying and deporting the 150,000 foreigners who overstay their legal visas each year. Of the thirty-nine thousand deported each year during the mid-1990s, only six hundred were ordered to leave for having overstayed their visas.

It seemed like an excellent idea to use motor-vehicle enforcement to identify and arrest those who were here illegally. But Clinton refused to pursue the idea. The idea ran into a solid wall of resistance led by White House adviser George Stephanopoulos. I pushed the proposal again at a meeting with Clinton on April 5, 1995, calling once more for "driver's licenses [to] expire when visas do." But no action was ever taken.

Stephanopoulos explained why in his 1999 memoir *All Too Human:*

Next on his [Morris's] list of potential presidential targets was immigrants. Basically, he wanted to create a background-check system that would turn your average traffic cop into a member of the U.S. Border Patrol. If, say, a police officer spotted a suspiciously brown-skinned person driving a car with a busted taillight, Dick's scheme would give him the ability to dial into a computer and order immediate deportation if the driver's papers weren't in order. Though he brushed off my fears of potential abuse and political harm to our Hispanic base, I persuaded him to hold off on the practical grounds of prohibitive cost.

The real story is a bit more complicated. White House deputy chief of staff Harold Ickes was charged with "vetting" the proposal through INS and the Justice Department. His answer was both decisive and shocking. "We can't deport the people we are already finding," he said. "If we expand the list of deportees without being able to act against them, the result would be a major scandal."

Even though I renewed the proposal at four subsequent meetings with the president, it was never adopted.

What a shame!

Three of the 9/11 hijackers had been pulled over by traffic cops in the months before 9/11. Had the drivers' license proposal been accepted, we might have sent them packing to the Middle East before they had their chance to fly airplanes into our buildings. In April 2001—five months before 9/11—Mohamed Atta, the ringleader of the hijackers, was stopped by police near Miami for driving without a license. He was summoned to appear in court, never showed up, a bench warrant was issued, and the matter ended. Had the motor vehicle/INS/FBI interface been functioning at that time, the traffic cop would have discovered that Atta was in the country illegally, his visa having expired in January 2001. Atta would have been arrested on the spot and bound over to the INS for deportation. He might not have been in the United States to lead the 9/11 hijackers on their grisly mission.

That same month, Nawaf Alhazmi, one of the hijackers who later seized control of American Airlines Flight 77 and crashed it into the Pentagon, got a ticket (in Oklahoma City, of all places) for speeding.

And Ziad Samir Jarrah, one of the four hijackers of United Airlines

Flight 93, the plane that crashed in Pennsylvania, was pulled over on September 9, just two days before the attacks, for driving between ninety and ninety-five miles per hour on Interstate 95. As CNN correspondent Jonathan Aiken observed, "Before 9/11 there really was no terrorist wanted list that a state trooper or anyone in the state agency could turn to indicate that there was any federal interest in this individual. As far as the state police in Maryland knew, Jarrah was a law-abiding citizen. . . ."

Had such a list existed, or had the INS and FBI been interfaced with the motor-vehicle computers, things might have been different.

Something always came before fighting terrorism. Some other policy or political consideration always had priority. In this case, Stephanopoulos was likely close to the mark when he warned of the harm to Clinton's "Hispanic base." Since the vast majority of illegal immigrants—although not terrorists—came from Mexico and other Hispanic countries, any program of this sort might be stereotyped as encouraging racial profiling by traffic cops.

To the Clinton White House, it was just more important to be friendly to Hispanic voters in the short term than to hasten deportations, and thus protect Americans of all races, in the longer term.

Terrorism Strikes: Oklahoma City Bombing

My first word of the bombing of the Murrah Federal Building in Oklahoma City came when Clinton called me while I was on vacation (in Paris again, as it happened). He was distraught, almost in shock: "Haven't you heard? Dozens, maybe even hundreds, of people were killed." At first Clinton thought the terrorists were from the Middle East, but it shortly became clear that the culprits were domestic fanatics.

In the face of such an assault, I urged Clinton to use an address to the nation to propose bold steps to counter terrorism. He demurred. There was always a reason. "If I do that the FBI says that I might bring on a second attack. I've got to move carefully here. This is a dangerous situation."

As Clinton confronted the Oklahoma City bombing, with its 168 deaths, it became increasingly clear that he was more comfortable offering America spiritual leadership in the struggle to find meaning in the piles of rubble than he was in taking practical steps to thwart future attacks.

Abandoning the role of commander in chief in favor of the soothing tones of a mourner in chief, he told the grieving relatives of the Oklahoma City dead: "Today our nation joins with you in grief." He urged Americans "to purge [themselves] of the dark forces that led to this evil." In a reprise of John F. Kennedy's cold war injunction that those who thought freedom was in retreat should "come to Berlin," he said: "If anybody thinks Americans are mostly mean and selfish, they ought to come to Oklahoma. If anybody thinks Americans lost their capacity for love and courage, they ought to come to Oklahoma." The New York *Daily News* reported that "the President told the victims' families, many weeping, that wounds take a long time to heal. 'But,' he added, the healing 'must begin.'"

Appearing on *60 Minutes,* Clinton stressed the emotional and spiritual implications of the Oklahoma City bombing, using his enormous capacity for empathy to ease the suffering of those who had lost loved ones and of a nation in shock. Reaching eloquently into the nation's soul, Clinton drew spiritual conclusions from the bombing and gave advice on how to handle the aftermath. "The anger you feel is valid but you must not allow yourselves to be consumed by it. The hurt you feel must not be allowed to turn into hate, but instead into the search for justice. The loss you feel must not paralyze your own lives. Instead, you must try to pay tribute to your loved ones by continuing to do all the things they left undone."

As I watched the coverage, I kept waiting for the other shoe to drop—for the president to make specific proposals to stop terrorism before it spread further. It seemed obvious that a climactic opportunity was being wasted. But each time we spoke, the president said he felt handicapped by the FBI and by the needs of the investigation.

He did say, as the *New York Times* put it, that he would "seek new authority for federal agents to monitor the telephone calls and check the credit, hotel, and travel records of suspected terrorists." Later, he supplemented this proposal with one to require that "taggants" be added to explosive materials to make them easier to track should they be used in terrorist attacks.

On the key issue, though, Clinton demurred. In the wake of Oklahoma City, the FBI asked for broader powers to investigate terrorist groups. As the *Times* explained on April 25, 1995:

> Under current guidelines, the FBI is forbidden from investigating [terrorist or extremist] groups unless there is a "reasonable indication" that they are trying to achieve their goals through violence and explicit violations of the criminal laws. Following the Oklahoma City bombing, law-enforcement officials have complained privately that those guidelines hamper them from gathering the kind of information needed to prevent such tragedies. Under the proposal being considered, the FBI could infiltrate such organizations or use informers to keep track of their activities.

But the proposal to expand FBI powers ran into opposition from the Treasury Department in general and the Bureau of Alcohol, Tobacco, and Firearms (ATF) in particular. Faced with a split in his own ranks, Clinton flinched; in the end, he never proposed altering the ground rules for surveillance that had been tying up the FBI.

In his antiterrorist package, the *Washington Post* noted, Clinton had refused to abandon "the requirement that law enforcement officials prove there is 'probable cause' of criminal activity before a judge approves surveillance against a suspect." The FBI had sought such powers "to compile information on potentially menacing organizations . . . even when there is no evidence they are involved in criminal activity."

Part of the problem in battling terror during the Clinton administration was the attorney general, Janet Reno. When President Clinton told me, in the summer of 1995, that her appointment was "his worst mistake," he was alluding to a long series of anticrime measures torpedoed by Reno, on grounds ranging from civil rights to civil liberties to budgetary constraints to pride of authorship.

Bill Gertz catalogs just one of the ways in which Reno undermined America's ability to prepare for 9/11. When the Minneapolis office of the FBI was alerted to the flight lessons being taken by Zacarias Moussaoui, alleged to be al Qaeda's twentieth hijacker, agents sought access to his laptop computer. But FBI headquarters denied the request, citing the lack of probable cause that a crime had been or was about to be committed. Commenting on the decision after 9/11, John L. Martin, a former Justice Department official, told Gertz he believed "that if the FBI had gone to the career lawyers in the . . . Internal Security Section of the Criminal Division in the Justice Department, they would have been advised to go after the laptop on any number of legal grounds."

Why didn't they?

Because in 1994, Reno's Justice Department adopted new rules that barred the FBI from contacting the Internal Security Section of the Justice Department, as Gertz explains, "as part of an effort to control FBI activities in the intelligence arena."

Whether the focus was deportation of illegal immigrants or expanding FBI investigative powers, the harm Reno did to American national security in the fight against terror was incalculable.

As Clinton's adviser, I chafed at the administration's lack of substantive measures in the wake of Oklahoma City, urging, in vain, stronger steps to counter and prevent terrorism. While many of these ideas were tailored to combat the domestic terrorists who had killed so wantonly in Oklahoma City, they would have done much to move America's war against terror, both foreign and domestic, into high gear.

In a White House meeting on April 27, 1995, two weeks after the attack, I called on Clinton to reject "the tombstone approach which only acts after terrorism has happened." I suggested "preventative surveillance and public disclosure of terrorist group activities to save lives before criminal actions are committed."

Specifically, I suggested that Clinton move to curtail charitable donations to groups funneling funds to terrorists. I suggested that he create a " 'President's List' of extremist/terrorist organizations to warn the public against well-intentioned donations which might foster terrorism." I urged "public disclosure of membership lists and donor lists by such organizations to aid in the investigative process."

The civil liberties crowd reacted with horror and rallied to persuade the president to kill the idea. Stephanopoulos recalls the play-by-play:

> Dick wanted a "national crusade" against domestic terrorism. In the wake of the Oklahoma City bombings, Morris didn't think that you could be too tough on the militias. While Dick's read of public sentiment was unassailable, his proposals reminded me of

the advice his late cousin Roy Cohn used to give Joe McCarthy. Morris wanted to require militia groups to register their guns and their membership with the FBI and he wanted the Justice Department to publish the names of suspected terrorists in the newspapers. I raised a civil liberties argument. "Oh, people don't care about that," he said. Then I countered with process, saying that if the attorney general wasn't on board (which she'd never be), Dick couldn't achieve his goal. Leaks from the Justice Department would only make Clinton look weak, and the paperwork would never emerge from the bowels of the bureaucracy unless the president typed it himself.

Stephanopoulos's comment is typical of how the White House staff sought to disempower the president. By threatening leaks and administrative noncooperation ("the paperwork would never emerge . . . unless the president typed it himself"!!!), they gleefully controlled this oft-weak chief executive. Can you imagine a White House staff member having the temerity to pull this kind of stuff on George W. Bush?

Clinton did issue an executive order in 1995 freezing the assets of twelve "foreign terrorist" organizations, mostly connected with the Palestinians. But it was only after al Qaeda's bombings of embassies in Kenya and Tanzania in 1998 that "the Clinton administration began the first major effort to disrupt the network's financing," as the *New York Times* reported, by freezing the assets of al Qaeda. But, since the terrorist group wasn't so obliging as to hold a bank account in its own name, the order netted nothing.

Little was done to enforce even the limited executive orders on terrorist fund-raising President Clinton had issued. Before 9/11, according to the *Washington Post*, "the number of cases brought under those orders can be counted on little more than one hand. Nearly all have involved individuals and organizations whose money was allegedly being funneled to the Palestinian groups Hezbollah and Hamas. None of it has related directly to bin Laden or al Qaeda."

The *New York Times* reported that "beginning in 1999, midlevel Clinton administration officials traveled to Saudi Arabia, Kuwait, Bahrain and the United Arab Emirates [U.A.E.] seeking information about charities aiding al Qaeda. But Saudi Arabia and the U.A.E. provided no assistance . . . with the embassy bombings receding into memory, the administration largely moved on."

Former Carter domestic adviser Stuart Eizenstat, a key figure in U.S.-European relations, noted that "these visits were not followed up by senior-level intervention by the State Department, or for that matter by Treasury, to those governments. I think that was interpreted by those governments as meaning this was not the highest priority." To put it mildly.

Clinton's diplomats might have done more than just ask the Saudis about charities that aided al Qaeda. According to a report by Jean-Charles Brisard, an investigator hired by the United Nations to report on al Qaeda funding sources, Saudi Arabia transferred $500 million during the 1990s to the terrorist gang. As Brisard puts it: "One must question the real ability and willingness of the kingdom to exercise any control over the use of religious money in and outside of the country."

It wasn't until after 9/11 that the U.S. government, under President Bush, finally closed down charities that funneled money to terrorist groups. When Bush did move against al Qaeda's terror-funding sources, he found an extensive network of back-channel funding and moved aggressively—as Clinton could and should have done—to disrupt it.

The *New York Times* reported that a key element in al Qaeda's cash flow came from "a financial network called Al Barakaat, which owns an . . . informal remittance system that moves millions of dollars around the world with virtually no paper trail." Immigrants use Al Barakaat "to send money back home . . . where terrorist operatives siphon off a portion of it for al Qaeda." The terrorists charge a 5 percent fee on each transfer, a kind of terror surcharge.

Bush froze Al Barakaat's operations and, through an executive order, expanded the president's authority to block assets of foreign entities that aid terrorism. Why didn't Clinton do that? He had the power, just as Bush did. He had the intelligence information. It was already clear how dangerous al Qaeda was. Why did he leave the time bomb ticking?

Only rarely did Clinton actually oppose any recommendations to fight terror or even make an affirmative decision to put other priorities first. There was never a meeting where Clinton listened to the ideas, cleared his throat, and announced his decision. That wasn't how the White House worked in the 1990s.

Instead, the president would hear ideas proposed in staff meetings or at our weekly political strategy sessions on Tuesday or Wednesday nights in the East Wing residence. He usually remained silent, declining to comment on the proposals under discussion. Burned by leaks early in his term, when he was more forthcoming with his reactions, Clinton told me, "I have learned not to say anything in front of more than one other person."

His silence nevertheless sent a clear message to me and the others involved with his policies and message—"check it out." Run the idea by the various cabinet departments and agency heads and see what they think about it. Vet it by the National Security Council or the economic team and get their reactions.

It was at this stage that the proposals to battle terrorism usually ran into trouble. Without a clear mandate to put terrorism at the top of the national agenda, every idea ran into opposition at some point in the bureaucratic food chain. The Bureau of Alcohol, Tobacco and Firearms killed the FBI's proposals to expand its powers of surveillance over possible terrorists. Clinton wouldn't move against fund-raising fronts for terrorist groups, because Attorney General Janet Reno refused to agree. The administration decided not to stop illegal aliens from getting drivers' licenses, because the Immigration and Naturalization Service had too big a backlog of deportation cases already. Something was always more important than fighting terrorism.

And when one of these proposals ran into bureaucratic opposition, Clinton just let it die. The slightest hint of disagreement from a law enforcement, civil rights, or military perspective was enough to send him scurrying for cover.

Clinton wasn't this way on every issue—not on the domestic front, for instance. Clinton ran roughshod over his own liberal staff at the White House and at the Department of Health and Human Services to sign the welfare reform bill. When his Department of Education and Office of Management and Budget people objected to his proposal to offer tax credits to offset tuition for the first two years of college, he demanded that the initiative proceed on schedule. His Budget Office objected to his decision to propose a balanced-budget plan, but Clinton did it anyway.

But on issues of terrorism, defense, and foreign affairs, generally, he was always too wary of criticism to act decisively. He was never strong enough to take the kinds of stands necessary to override the stand-pat instincts of his bureaucracy.

Even those initiatives the president did take after Oklahoma City ran into opposition from the Republican-controlled Congress. Sometimes their dissent was politically motivated—the GOP was always eager to deny the president an achievement on which he could run for a second term—but often it was based on their desire to protect the right wing, the GOP political base.

After Oklahoma City, terrorism was seen primarily as a threat from the extreme right wing. The skinheads, militiamen, gun nuts, white supremacists, and anarchist-libertarians of the radical right were portrayed as a subculture that gave rise to the Oklahoma City bombing, and it was against them that most national angst about terror was directed.

Immediately after the bombing in April 1995, congressional Republicans, who controlled both Houses, vowed to pass antiterror legislation within six weeks. Congress did, indeed, pass antiterror legislation in April—but not until April 1996, a full year after the bombing. Fearful of being lynched in public for failing to pass antiterror laws while America mourned the first anniversary of the bombing, Congress sent the president a watered-down bill that deleted his two most important proposals: expanded wiretap authority and the use of taggants to identify bombs.

His request for more wiretaps was omitted from the bill entirely. Clinton had proposed that federal wiretap procedures be revised so law-enforcement officials could "follow terrorists as they move from phone to phone." Under this measure, a warrant would allow agents to tap all phones the suspect used—cellular, wireless, in-home, or pay phones—rather than just one specific instrument. As Clinton noted, "This authority has already been granted to our law-enforcement officials when they're dealing with organized criminals," but Congress refused to allow its use in the fight against terror.

Clinton proposed that taggants ("trace chemical or . . . microplastic chips") be added to all possible sources of explosives (such as fertilizer), scattered throughout to permit, as he explained, "sophisticated machines [to] find bombs before they explode, and when they do explode, [to allow] police scientists [to] trace a bomb back to the people who actually sold the explosive materials that led to the bomb." As Clinton noted, taggants had been used in "Switzerland over the past decade [and have] helped to identify who made bombs and explosives in over 500 cases. When it was being tested in our country several years ago, it helped police to find a murderer in Maryland." Yet pressure from the National Rifle Association (NRA) eventually overcame the proposal; taggants did not make it into the final terrorism bill. The strange-bedfellows alliance of civil liberties groups and Republicans sympathetic to right-wing groups had killed both the wiretap authority and the taggants proposal.

The price America was to pay for kowtowing to the NRA became fully apparent when a pipe bomb ripped through the Atlanta Olympic Centennial Park in the summer of 1996. To date, authorities have not solved this crime. But, as Clinton pointed out in a radio address after the explosion, taggant technology well might have helped law-enforcement officials to trace the bomb.

Even after the bombing, all Clinton dared to ask of the Republican Congress was to conduct the study of taggants they had authorized, but not funded, in the previous round of antiterror legislation, and to ask that it be extended to study the safety of taggants in black or smokeless gunpowder. The Republicans wouldn't consider even this; two months after the Atlanta bombing, they tabled a Democratic amendment to fund the study, by a vote of 57–42.

The strange story of how the Clinton antiterror bill was gutted in Congress was laid out fully in a *Washington Post* article by Lally Weymouth on August 14, 1996. Noting that the legislation was emasculated by "a bizarre coalition dominated by the far left and the extreme right," she explained how the bill was "watered down [to deny] law-enforcement authorities tools needed effectively to combat the growing terrorism menace."

At the core of this "profoundly strange alliance" was a coalition between "the GOP's far right—led by Representative Bob Barr of Georgia—and the Democratic far left—mobilized by Represenstative John Conyers of Michigan."

Would the provisions for taggants and wiretaps have prevented the Atlanta bombing? Would they have led law-enforcement agents to the front door of those responsible? We will never know.

But we do know that President George W. Bush felt that the wiretap provisions in the Clinton bill were so important that they formed a key part of the post-9/11 antiterror bill he pushed through Congress. One cannot help but wonder: If the Republicans had been less blind and the Democratic left less self-destructive, would federal law enforcement have been more effective in preventing 9/11?

Nevertheless, the antiterrorism bill made a great photo opportunity for politicians of both parties. Twenty-two members of Congress gathered on the South Lawn of the White House for the bill signing, including Clinton's future opponent in the 1996 election, Senate Majority Leader Bob Dole.

Unfortunately, the substance of the bill was totally inadequate, especially in light of the 9/11 experience. Here's what it provided:

- Authorized $1 billion over four years to help federal officials, especially the FBI, monitor and catch terrorists. (*Big deal: In the rush to catch up after 9/11, Bush had to increase antiterror spending by tens of billions in one year to make a difference after the true dimensions of the challenge became clear.*)

- Restricted habeas corpus petitions by state and federal inmates and curtailed the power of U.S. judges to overturn convictions in state courts. (*This was a rider attached to the bill by the Republicans; it was the only way they could get Clinton to sign it into law. It had nothing to do with terrorism.*)

- Made foreign airlines using U.S. airports adopt, as the Cleveland *Plain Dealer* reported, "the same stringent security measures as U.S. carriers." (*These measures were not so "stringent" as to stop the 9/11 hijackings.*)

- Tagged plastic explosives to make it easier to track bombs. (*This watered-down version of Clinton's taggants proposal turned out to be useless in tracing the Atlanta Olympic bomber.*)

- Banned fund-raising in the United States by foreign terrorist groups, as designated by the secretary of state. (*Which Clinton had already done by executive order.*)

- Banned financial transactions between Americans and terrorist states like Libya, Syria, and Iran. (*This provision was, of course, of no use in fighting terrorist groups that were not nations, such as al Qaeda.)*

- Allowed Washington to deny visas for foreigners suspected of belonging to terrorist groups. *(This provision sounds very good in the wake of 9/11, but it really amounts to very little. U.S. intelligence in terror-sponsoring nations is too limited—and diplomatic presence there is usually nonexistent—to permit us to identify who is dangerous and who is not.)*

- Permitted faster deportation of foreigners convicted of crimes while in the United States. *(While this measure makes sense, it has nothing to do with fighting terror. Robert Mueller, director of the FBI, said of the 9/11 hijackers: "While here [in the United States], the hijackers effectively operated without suspicion, triggering nothing that alerted law enforcement and doing nothing that exposed them to domestic coverage.... They committed no crimes, with the exception of minor traffic violations. They dressed and acted like Americans, shopping and eating at places like Wal-Mart and Pizza Hut.")*

- Made it a crime to use chemical weapons in the United States or against our citizens abroad. *(Sure to send terrorists into a panic!)*

Congress had labored for a year on its antiterror package, but all it produced was this paltry list of half measures. The very limited nature of its scope and reach reflects, eloquently, the low priority terrorism received in official Washington in the middle of 1996.

The Pseudo-Sanctions Against Iran

If the Republicans were loath to approve antiterrorism measures at home that might annoy their more extreme right-wing supporters, they were determined to force Clinton's hand and make him take bold action against terrorists in the Middle East. Senator Alfonse D'Amato (R-N.Y.) took the lead in 1995 by introducing legislation extending the U.S. oil embargo against Iran to limit the ability of foreign companies to assist Teheran in developing its oil and gas industry.

Clinton had imposed the embargo blocking U.S. companies from helping Iran's petroleum industry in the spring of 1995, after Secretary of State Warren Christopher was enraged by the decision of Conoco, a Du Pont company, to make a deal with Iran. The *New York Times* reported that Christopher "argued that the United States should take the lead in depriving Iran, an outlaw country, of the financial resources it needed to develop nuclear weapons or sponsor terrorist activities."

Before the embargo, U.S. companies had been investing more than $4 billion annually in Iran's oil industry. The American embargo, of course, cut off these deals, but European companies continued to do business with Iran. Senator D'Amato was anxious to stop foreign companies, whose nations did not honor the embargo, from taking up the slack and helping Iran earn more from its oil reserves. Spurred by the decision of Total S.A., a French oil refiner, to take over the Conoco deal, D'Amato's legislative proposals imposed sanctions against any foreign company that aided Iran's oil and gas industry. The penalties included a ban on the importation of their products into the United States, and a prohibition against loans to the company by any U.S. bank. The Federal Reserve Board would also be directed to bar any financial institution from becoming a primary dealer in bonds of U.S. origin if they had aided energy projects in Iran.

At first the administration dismissed the D'Amato bill as just partisan posturing, introduced to allow the New York senator to strut in front of his large, domestic Jewish community. But soon the legislation gained momentum, and Clinton was forced to take it seriously.

Angered at American attempts to block European companies from involvement in lucrative deals with Iran, the European Union (EU) blasted the legislation, insisting, as the *Times* reported, that the United States had "no basis in international law to claim the right to impose sanctions on any foreign person or foreign-owned company who supplies Iran with oil development equipment."

Clinton felt whipsawed by the conflicting pressures on the D'Amato bill. It was gaining momentum in the Senate, spurred by fear of Iranian terrorism, but the European Union was threatening to appeal to the World Trade Organization (WTO) if the bill passed. Clinton partially solved the problem by getting the Republicans to water down the legislation, dropping the crucial provision banning imports of all products made by companies doing business with Iran. The president hoped that this would cool European anger at the bill. The Senate passed the legislation in December 1995.

But Europe was still unhappy. Within the administration, Deputy National Security Advisor Sandy Berger urged Clinton to oppose the legislation unless it included a provision permitting the president to waive the sanctions when he considered it in the "national interest." But the Republicans pressed hard for passage to impose broader sanctions against Iran.

Responding to European concerns and the cautious advice of his own National Security team, Clinton insisted on the national security waiver as the price for his signature on the bill. Complying with White House pressure, the House passed the watered-down legislation on June 20, 1996.

Europe still went ballistic, however, threatening retaliation if the sanctions were ever imposed on their companies. Germany was particularly sensitive. Anxious to assure Iranian repayment of its $8.6 billion debt, Berlin had been alarmed by the drop in its exports to Iran from $5.2 billion in 1992 to only $1.6 billion in 1995. Claiming that Clinton was only grandstanding before a domestic political audience, German foreign minister Klaus Kinkel said that it was "better to continue the dialogue with Iran rather than break off all contacts, introduce sanctions, and further radicalize Iran by isolating the country."

The Germans were right about one thing. Clinton was grandstanding when he signed the Iranian sanctions bill on August 6, 1996, days before the Republican National Convention nominated Bob Dole as his opponent. Piously, the president told our allies, "you cannot do business with countries that practice commerce with you by day while funding or protecting the terrorists who kill you and your innocent civilians by night."

But what Clinton didn't say when he signed the bill was that he never planned to enforce it. For the rest of his presidency, whenever a European company tripped the wire that should have led to sanctions, Clinton demurred, invoking the national security waiver. Hypocritical in the extreme, he had made a show of his toughness by signing a bill he never intended to use and by approving sanctions he never planned to impose.

Clinton may not have used the new law to stop Iran's oil industry from bankrolling terror, but the legislation certainly helped enliven his acceptance speech to the Democratic National Convention on August 29, 1996. In a speech that was breathtaking in its hypocrisy (in light of Clinton's subsequent willingness to waive sanctions against companies trading with Iran), the president told the convention: ". . . We are working to rally a world coalition with zero tolerance for terrorism. Just this month I signed a law imposing harsh sanctions on foreign companies that invest in key sectors of the Iranian and Libyan economies. As long as Iran trains, supports, and protects terrorists, as long as Libya refuses to give up the people who blew up Pan Am 103, they will pay a price from the United States." *(Applause)*

Some price! Not a single company lost a single dollar, euro, franc, mark, pound, lira, peso, or yen as a result of U.S. sanctions against its investments in Iranian oil or gas fields. Not *one*.

If Americans were deceived by Clinton's posturing, Europeans weren't. As Clinton was signing the sanctions law, *USA Today* reported that "France, Germany and Britain, as well as the European Union, are among those who have threatened retaliation. But the hope in Europe is that 'after the elections, this law will be shelved or watered down,' says Steven Englander, economist with the Paris office of Smith Barney brokerage."

The Europeans had that right. The first real test of the new sanctions came in the fall of 1997, when the *Washington Post* reported that "French, Russian, and Malaysian oil companies . . . triggered a State Department investigation of whether they should be penalized under U.S. law for . . . developing a major offshore natural gas field in Iran."

Iran had been after capital to develop the gas field. The *Post* reported that "the Iranians scored their first major success last summer when Total S.A. of France, the giant Russian natural gas company Gazprom and the state-owned Petronas of Malaysia signed a contract to invest $2 billion in developing a gas field known as South Pars."

Six months later, the newspaper related how "Clinton's senior foreign policy advisers met late into the night . . . grappling with what might have seemed a straightforward decision: whether to impose legally mandated sanctions. . . . The administration appears paralyzed by the myriad arguments for and against sanctions."

The arguments were familiar. The supporters of sanctions said that the money from the gas field would go right into funding terrorism, while opponents worried that imposing them would injure NATO and hurt reformist forces in Iran. The *Post* reported that "according to some officials, the administration is basically content to postpone a decision because delay avoids potential negative consequences of a decision either way, while leaving the deterrent effect of U.S. sanctions hanging over other foreign companies."

Al D'Amato, the sponsor of the sanctions, warned Clinton that "if the United States does not take swift, decisive action to apply these available sanctions, we will have undercut our long-standing policy against Iranian terrorism."

In May 1998, Clinton caved in and waived the sanctions against foreign oil companies over the Iranian gas fields deal. All the president's strong words when he signed the sanctions bill in August 1996, which he repeated later that month at the Democratic National Convention, went up in smoke. When the challenge finally surfaced, Clinton ran for cover. Despite congressional action and his own commitments, sanctions were dead.

Bill Safire said it best in the *New York Times:* Dual containment against Iraq and Iran had been replaced by a "dual doormat" theory.

The Terror Summer of 1996

Sometimes, defenders of Clinton's record on terrorism plead that the national mood during his presidency was not sufficiently alert to the danger of attacks on our shores to permit him to take bold action. Certainly, there was never any real understanding of the magnitude of what could happen. Only a very few of the farsighted (such as former senators Gary Hart and Warren Rudman) could envision an attack of the severity of the 9/11 assault. But the national thermometer rose fairly high in the summer of 1996, as the focus on terrorism reached its greatest intensity. Three attacks, coming in close succession, attracted national attention and opened the political possibility of bold military action against foreign terrorists:

- On June 25, 1996, a bomb ripped through the Khobar Towers barracks in Dhahran, Saudi Arabia, that housed hundreds of U.S. airmen. The explosion left an eighty-foot crater; nineteen died and hundreds were injured.

- Three weeks later, on July 18, 1996, TWA Flight 800 exploded in midair and crashed into the Atlantic about sixty miles east of New York City, thirty minutes after taking off from Kennedy Airport. All 230 passengers died.

- On July 27, 1996, just ten days after the TWA crash, a bomb exploded in Centennial Park, the center of the Olympic Games under way in Atlanta, Georgia. The blast killed 2 people and injured 111 others. It shocked a nation that had been following the games avidly on television.

Hindsight has dulled the memories of that difficult summer of 1996. The Khobar Towers barracks bombing was, of course, the work of al Qaeda. The Atlanta bombing was seen, at the time and since, as the likely work of domestic terrorists. While the cause of the TWA crash has never been finally determined, at the time it was widely believed to have been a terrorist incident. On July 19, 1996, the *Boston Globe* reported that terrorism was "the operating theory behind the FBI's investigation of the crash of TWA flight 800."

But the nation drew no distinctions among the three attacks, lumping them together under one broad heading: terrorism. Americans demanded action. But all they got from Clinton were speeches.

In this atmosphere, there began to assemble a critical mass of public opinion for a truly aggressive strategy, lifting antiterrorism to the top of the nation's political agenda. Had Clinton responded more vigorously, and used the national mood for more aggressive action against terror, 9/11 might never have happened.

Inaction on the Khobar Towers Bombing

When a truck carrying the equivalent of twenty thousand pounds of TNT exploded outside the Khobar Towers barracks in June 1996, President Clinton had his usual stern words for the attackers: "The explosion appears to be the work of terrorists, and if that is the case, like all Americans, I am outraged by it. The cowards who committed this murderous act must not go unpunished."

And yet, when the Saudi Arabian government discouraged FBI director Louis Freeh's efforts to investigate the attack, Clinton acquiesced.

The Khobar Towers bombing was bin Laden's second attack in eight months in Saudi Arabia. On November 12, 1995, he had orchestrated a bombing of the Office of the Program Manager of the Saudi National Guard in Riyadh, Saudi Arabia. The U.S. military had used the building to train Saudi troops; five Americans were killed in the bombing. The Saudi government promptly arrested and quickly executed four men blamed for the attack. U.S. officials were never permitted to interrogate the suspects.

After the Khobar Towers attack, the *Washington Post* reported that "U.S. officials . . . suspect a link between the two bomb blasts." It mentioned that the Saudi government had "undermined" American efforts to "gauge the full scope of the threat to American military forces in Saudi Arabia" by its "reluctance to cooperate fully with U.S. investigators and intelligence analysts."

The dead men who the Saudis executed after the Riyadh attack told no tales.

For their part, the Saudis, according to the *Post,* "denied any friction between U.S. and Saudi investigators and said there was 'total cooperation' in the Riyadh bombing probe." The *Post* article, however, told a very different story. "Saudi security officials held one of those eventually convicted of the Riyadh bombing for three months, and the other three for one month, before they informed any officials at the U.S. Embassy there. The Saudi government was 'adamant about not letting us in there' to interview the suspects before they were executed, the official said." The newspaper quoted a senior U.S. law-enforcement official as saying: "They [the Saudis] didn't let the FBI interview these guys and then they killed them." The official speculated that the Saudis did not want the United States to interview the bombers because it was "fearful of what we might find out once the United States gets a complete picture of those connected to the Riyadh bombing or to dissident movements."

As we now know, that trail would have led straight to Osama bin Laden.

Understandably, U.S. law-enforcement officials were worried that their leads in the Khobar barracks bombing would be cut short by the Saudi's busy executioner. They were right.

The *Washington Post* reported that FBI director Louis J. Freeh traveled three times to the kingdom to "seek U.S. access to several individuals who have been detained by the Saudi government on suspicion of involvement in the Dhahran bombing." Chafing at the lack of Saudi cooperation, the paper quoted U.S. defense secretary William J. Perry as saying: "We cannot accept the problems we had the last time."

Despite U.S. entreaties, Assistant FBI Director Robert Brant told a congressional committee that "the Saudi Arabian government has prevented FBI investigators from interviewing any civilians who witnessed or may have been involved in the bombing."

His boss, Louis Freeh, was more diplomatic in his testimony: "We have not gotten everything we have asked for and this has affected our ability to make findings or conclusions or to channel the investigation in different directions. There is a great deal of information we have not seen."

It was not until June 21, 2001—five years after the bombing, and well into the Bush administration, that the United States indicted thirteen Saudis and one Lebanese for the bombing of Khobar Towers.

Why did Clinton permit the Saudis to drag their feet in cooperating with the investigation? Why was not more pressure put on our so-called allies to be forthcoming with their witnesses and evidence? The former president's failure to be more aggressive in pushing the Saudis ranks as a key intelligence failure.

A glimmer of what we might have learned had Clinton pursued the issue came in a 1997 CNN story headlined "Wealthy Saudi May Have Had Role in Khobar Bombing; An Investigation Is Under Way." Introducing bin Laden to the American public as an "elusive Saudi dissident," the network noted that "a criminal investigation being conducted by the U.S. attorney in New York City turned up two bin Ladin statements to newspapers and to CNN calling for a holy war against U.S. troops stationed in Saudi Arabia."

We will never know what Bill Clinton might have learned, had he put finding the terrorists who killed nineteen American servicemen ahead of smoothing Saudi Arabia's diplomatic feathers as a diplomatic priority.

More of Same: Olympic Bombing of 1996

At one point, as the summer began, it seemed as if President Clinton had gotten the point. Traveling to the fourteenth-century French town of Pérouges to meet with the G-7 world leaders, Clinton moved terrorism to the top of the agenda. Calling the need to fight terror "one of the great burdens of the modern world," Clinton got the leaders to declare, "We consider the fight against terrorism to be our absolute priority."

The *Boston Globe* of June 28, 1996, rhapsodized that "by successfully pushing his terrorism proposals, Clinton dominated the early agenda of the three-day summit, relegating many of the anticipated complaints over U.S. trade policies to secondary status."

After the Olympic bombing, Clinton seemed determined to take action. "We will spare no efforts to find out who was responsible for this murderous act," he said. "We will track them down, we will bring them to justice, we will see that they are punished."

But, in fact, Clinton did almost nothing to give effect to his words. He just dusted off his old proposals for taggants and wiretap authority and sent them to Congress.

It wasn't for lack of national consensus that Clinton acted so timidly after the terrorist attacks of the summer of 1996. His polling reflected a tremendous national focus on terrorism and its dangers. In a survey conducted for the president on June 6, 1995, before any of the three terrorist attacks, voters rated the battle against terrorism as our top foreign-policy issue, with 92 percent saying it was very important.

But after the trio of terrorist tragedy had struck in the summer of 1996, the national outcry grew. The president's poll of August 1 reflected the mood of tension and the desire for bold action. Asked if they would approve of "military action against suspected terrorist installations in nations that harbor terrorists or assist terrorists even if they didn't explicitly sponsor a terrorist act"? Voters backed action by 77–21.

By 84–14, they supported expanded wiretap powers, and by 77–19, they wanted the military to be involved "domestically and abroad to pursue terrorists."

The mandate for action was clear. The administration response was not.

The Air-Safety Debacle

Particularly in the area of air safety—after the TWA 800 crash—the public clamored for effective action. The history of attacks on passenger aircraft was prolific to anyone who sought to examine it.

- A year and a half before, in December 1994, Iraqi national Ramzi Yousef had admitted to detonating a bomb aboard Philippine Airlines Flight 434, ripping a two-foot-square hole in the fuselage while the plane was flying from Manila to Tokyo. After an emergency landing, one passenger died, and ten were injured.

- In December 1988, a Pan Am flight crashed over Lockerbie, Scotland, killing all 259 on board. Two Libyan terrorists have been convicted of the attack.

- On November 29, 1987, a North Korean agent planted a bomb on a Korean Airlines flight from Baghdad to Bangkok, killing all 115 on board.

- On April 2, 1986, a woman carrying a Lebanese passport, acting on behalf of a Palestinian terrorist, brought a bomb onto a TWA flight from Rome to Athens, killing four Americans, who were sucked through the aperture, and injuring nine others.

- On June 22, 1985, an Air India plane flying from Toronto to Bombay blew up near Ireland, killing all 329 passengers. The bomb that brought down the flight had been planted by Sikh extremists.

With so ghastly a history of air terrorism, public demand for greater protection in the skies escalated rapidly after the TWA crash. A survey I conducted for the president on July 24, 1996, indicated strong public support for dramatic measures to counter aircraft hijacking and bombing.

By 90–7, voters backed "modern X-ray machines at airports to examine all checked luggage."

By 92–6, they supported federalizing security personnel who worked at American airports.

By 92–6, they backed requiring photo identification for all air passengers.

Does this list of measures sound familiar? None were implemented by the Clinton administration, despite such broad public support. But each became public policy in the wake of the 9/11 hijackings. Unfortunately, none were in effect early enough to have prevented the attacks in the first place.

Instead of taking the bold actions suggested by some of his advisers, though, the president simply punted. He appointed a Commission on Aviation Safety and Security, to be headed by Vice President Al Gore, to report on steps to improve air safety (after the election, and after the furor had died down). Clinton raised high hopes for the commission in his acceptance speech to the Democratic Convention, loudly proclaiming: "We will improve airport and air travel safety. I have asked the vice president to establish a commission to report back to me on ways to do this. But now we will install the most sophisticated bomb-detection equipment in all our major airports. We will search every airplane flying to or from America from another nation—every flight, every cargo hold, every cabin, every time." *(Applause)*

What really happened after these far-reaching promises? Not much. After receiving Gore's report, on September 10, 1996, Clinton proposed to spend $429 million to improve security at U.S. airports as part of a $1.1 billion plan to fight terrorism worldwide.

USA Today reported, "The plan includes provisions to increase the number of federal agents guarding against terrorism, equip the nation's airports with high-tech bomb detection devices and track by computer passengers with suspicious travel patterns." Clinton also ordered immediate implementation of a requirement that all bags be matched to passengers on an airplane as a precondition of takeoff, a measure that reflected the happy assumption that no terrorist would ever choose to commit suicide. Clinton also ordered criminal background checks of airline workers and the deployment of bomb-sniffing dogs at key airports.

Based on these totally inadequate measures, Clinton predicted that "not only will the American people feel safer, they will be safer."

Nowhere in Gore's recommendations or in Clinton's proposals were the key measures his advisers had recommended and polls indicated voters approved: federalization of air-safety workers, photo identification for air travelers, and X-ray examination of all checked baggage.

In fact, Clinton immediately ran into trouble on the only really important part of his air-safety program. Buried in its text was a recommendation to spend $10 million on "automated passenger profiling." The administration said the program would be "based on information that is already in computer data bases." It would separate "passengers who present little or no risk and a small minority who merit additional attention."

The *New York Daily News* noted, "The commission was intentionally vague on which passengers would get extra eyeballing, but said that travel histories spun out by the computers might trigger alarms if the passenger showed frequent flier miles to Iran or Libya."

The left was outraged. "Rounding up the usual suspects may have been okay in *Casablanca,* but it's not okay in America," said Gregory Nojeim, legislative counsel for the ACLU.

The ACLU might have spared themselves the trouble of issuing the statement. Gore's recommendations for profiling didn't amount to much. Ultimately, the airlines voluntarily implemented their own system to decide who was a risk and who should get extra attention. The FBI and other law-enforcement agencies objected that the system wouldn't work and that it was based on far too limited data to be effective.

The Gore commission's final report, published in February 1997, was timid, its recommendations quite limited. *USA Today* commented that "Vice President Gore's Aviation Safety and Security Commission had the opportunity to effect dramatic reforms making U.S. flying safer. Unfortunately, the commission opted for a slow flight and an uncertain landing."

Forever addicted to hyperbole, Clinton said he would use "all the tools of modern science" to make air travel safe. Unfortunately, he failed to use even basic political science to get even the limited recommendations of the Gore commission approved. The commission's prediction—that its proposals would cut aviation disasters by 80 percent over the next ten years—is laughable in the aftermath of 9/11.

USA Today noted the holes in the commission report: "The commission instructed the Federal Aviation Administration to mandate security upgrades such as installing more explosive detection devices at airports, but it didn't say when. The commission recognized the critical need for criminal background and FBI fingerprint checks for airport and airline employees with access to secure areas. Yet it gave the airlines until mid-1999 to do so."

The commission recommended a similarly leisurely schedule for implementing the requirement that checked bags be matched with passengers on the plane. In its preelection-day report, Gore had recommended bag match testing within sixty days. Now he approved a delay until the end of 1997 before starting the plan and set no deadline for total compliance.

Gore's stress on bag matching was a good example of entering a new challenge perfectly prepared to meet the old one. The 1988 explosion that brought down Pan Am Flight 103 over Scotland had been traced to an extra passenger-less bag, with explosives.

But the idea that bag matching would be effective, in a world of suicide/homicide bombers, is itself ridiculous, indicative of the stultified thinking of the Clinton/Gore era. Gore did not even recommend fire suppression or smoke detection systems in the cargo holds of passenger airlines, the shortcoming that contributed to the death of 110 people aboard ValuJet Flight 592, which crashed in Florida in May 1996.

But even the limited steps recommended in the Gore report were watered down in Congress. Mark Green, in his book *Selling Out*, documents the efforts of the Air Transport Association, the lobbying arm of the airlines, to dilute, delay, or dismember the Gore recommendations.

> The ATA has used extensive lobbying and contributing to delay Congress from enacting the suggested requirements. In 2000, it lobbied to weaken legislation that would have mandated background checks for all airport screeners. That year, the top nine airlines plus the ATA spent $16.6 million on lobbyists, ten of them former members of Congress, two of them former secretaries of the Department of Transportation, which oversees the FAA, and another three former senior officers at the FAA. There were 210 lobbyists in all, and with their help the industry was successful in curbing new regulations.

When these same airlines now plead for help in the face of declining air travel after 9/11, they should be ashamed of their opposition to safety and antiterrorism measures in the 1990s, and realize how shortsighted and self-destructive their positions were.

The magnitude of the missed opportunities during the summer of

1996 cannot be exaggerated. The critical mass of public opinion and outrage was there to permit real action on air safety and terrorism. But Clinton and Gore—and the airlines themselves—blew it.

As a result, even if Bush and Cheney had realized the magnitude of the threat America faced in the months before 9/11, there was no way they could have acted effectively to keep the hijackers off the airplanes. With no system in place to check the identity of those traveling, no special training for security screeners, and no requirement for early boarding to allow time for thorough body searches, there was nothing they could have done to stop the hijackings. That fight was lost in 1996, when Clinton and Gore failed to act.

Iraq: Saddam Plays Clinton

When George H. W. Bush handed the White House over to Bill Clinton, Saddam Hussein was as completely under wraps as it is possible for a foreign leader of a sovereign state to be. His nation was blocked from selling oil and swarming with U.N. inspectors. Without revenue or the privacy in which to rearm, Saddam and his shattered military posed little international threat.

But when Clinton passed power to Bush's son eight years later, Iraq was frantically rearming, its coffers bulging with $40–$60 million income *daily* from the sale of 2 million barrels of oil. Arms inspectors were nowhere to be found, having been thrown out of the country by the Iraqi dictator. Saddam was building a war machine that would once again frighten the world with its potential for deadly weapons of mass destruction.

How did Saddam get Clinton to let him off the mat? It was like taking candy from a baby.

Each year brought a new demand from Saddam Hussein—to loosen sanctions, increase his oil revenues, curb inspectors, and, eventually, restore his complete freedom of action. His pattern repeated itself like a knitting stitch—back one, forward two.

First, Saddam would announce that he was going to refuse to honor some aspect of his agreement with the United Nations, cemented amid the ashes of his utter rout in the Gulf War of 1990–1991. Then the world would convulse in crisis. Emergency negotiations would ensue;

Saddam's allies—France and Russia—would press for concessions. Clinton would rattle his saber by bombing or sending troops to the Gulf. Then Saddam would seem to give in to U.S. demands. American foreign-policy officials would deny that they had made any concessions, and Clinton would take the bows for standing up to Saddam. Then, quietly, after the world's attention had shifted, the United States and the United Nations would grant some concession to Saddam as the previously agreed price for his keeping his past promises, all the while denying that they were ceding anything.

As long as Saddam was willing to be "humiliated" before the American public and let Clinton play the part of the tough and resolute president on the public stage, he could get away with anything—and eventually did.

Just three months after his inauguration, Iraq began to test Clinton's resolve. It plotted to kill former President Bush by exploding a car bomb during his postpresidential visit to Kuwait on April 14–April 16, 1993. After two months of investigation, Clinton determined that the plot had been orchestrated by Iraq. On June 27, 1993, Clinton dispatched twenty-three Tomahawk missiles to attack the Iraqi intelligence headquarters where the plot had been hatched. Calling his response "firm and commensurate" with the offense, Clinton told the nation, "We will combat terrorism. We will deter aggression. We will protect our people."

Like schoolmates sizing each other up on opening day, the bully Saddam took his measure of the ingenue Clinton—he would bomb but not invade. Bombing, Saddam could take. You can't get removed from power by bombing, absent the unlucky hit.

Saddam had two problems when Clinton took office: He needed to lift the embargo on the sale of his oil, and he had to get rid of the U.N. inspectors so he could spend the proceeds on arms rather than on food. Throughout the Clinton administration, Saddam worked first on one end of his problem, then on the other, like a man flexing first one wrist and then the other to loosen the ropes that bind him.

Saddam began by persuading Turkey to sell 12 million barrels of Iraqi oil stuck in a pipeline on its territory, resulting in $120 million in revenues to Saddam.

While noting that the Turkish plan "does have some elements" that might violate the strict ban on the sale of Iraqi oil, Western diplomats let the sale go through while Secretary of State Warren Christopher "expressed U.S. determination to resist any easing of U.N. sanctions."

As Iraq showed the world photos of its starving children, liberal and humanitarian pressure grew for easing of the sanctions throughout Clinton's first term. Saddam used the 20 million Iraqis, suffering under his boot, to strengthen his case to let him sell his oil.

When the United States and the United Nations would offer to permit oil sales under strict controls, Saddam would refuse, denouncing it as a violation of his national sovereignty. Aware that international pressure to drop the sanctions would grow as long as he let his people starve, Hussein held out for terms that would permit him to divert the oil revenue to rearmament.

Meanwhile, France and Russia demanded an end to all sanctions against Iraq. That put Clinton in the position of pushing to allow Iraq to sell a limited amount of oil to buy food, as an alternative to ending the sanctions.

When Saddam rejected two U.S.-British proposals in 1995–1996 to let him sell his oil under strict controls, the Iraqi dictator turned them down as an "insult" to his country's sovereignty. That sent U.N. Secretary-General Boutros Boutros-Ghali scurrying to negotiate with Saddam, seeking a way to start a flow of humanitarian aid, while the United States and the United Kingdom made a show of vigilantly scrutinizing potential deals so that Iraq could not "manipulate the oil sales agreement for its own ends," in the words of the *New York Times*.

So Saddam had manipulated the world into pressing him to agree to sell his oil, under a regimen that would control his use of the money, to assure that it went for food for his people. At first, Baghdad seemed to resist U.S. and British plans for restrictions on the oil-for-food program, yielding only reluctantly to international pressure for strict controls.

In reporting the deal, the *Washington Post* noted that "Iraq must accept stringent U.N. monitoring to ensure that the money is not used to buy weapons, luxury goods, or other items of benefit to Saddam's regime. . . . In particular, the United States and its allies insisted successfully on U.N. supervision of the banking arrangements for oil sales, minute U.N. scrutiny of how humanitarian supplies are to be distributed, U.N. control over delivery of aid to the breakaway Kurds in northern Iraq, and widespread discretionary power for U.N. monitors."

But Saddam had already achieved the biggest part of his goal. He could sell his oil. Now he set to work on the other half of his agenda: circumventing the limitations on what he did with the money.

The so-called controls were a sham from the beginning. Iraq was allowed to sell seven hundred thousand barrels of oil daily, a total that ultimately swelled to almost 2 million (two-thirds of its pre–Gulf War total). In return, Saddam had to abide by only the loosest of actual controls over the Iraqi use of the funds it generated. The restrictions the Clinton administration negotiated largely related to peripheral aspects of the deal, rather than to the core issue of preventing the use of the bulk of the money for restoration of Iraq's military and Saddam's regime.

The safeguards included letting the United Nations choose the bank that would handle the oil transactions and reliance on U.N. statistics in determining priorities for the distribution of the aid.

But Saddam realized, as Clinton apparently did not, that oil is fungible. Once the restriction on selling Iraqi oil was lifted, nobody could be sure that the oil a nation used was "legal" (i.e., allowed under oil-for-food) or "illegal" (i.e., smuggled) oil. It all looked the same—black.

Senator Frank Murkowski, the Republican senator from Alaska who chaired the Committee on Energy and Natural Resources, explained in 1999 how Iraqi oil ended up as arms for Saddam. "Illegally sold oil is moving by truck across the Turkish-Iraqi border. A more significant amount is moving by sea through the Persian Gulf. Exports of contraband Iraqi oil through the Gulf have jumped some fiftyfold in the past two years to nearly half a billion dollars. Further, Iraq has been steadily increasing illegal exports of oil to Jordan and Turkey."

Absurdly, the national media interpreted Iraq's willingness to accept these weak restrictions on the oil sales program as evidence of restiveness among its 20 million people. Saddam must have been feeling the heat from his starving millions at home, the media explained. But Bob Dole, Clinton's 1996 adversary, had it right. The deal gave Saddam "a source of revenue" with which to continue "his reign of terror." Piously, the administration rejected Dole's criticism, saying that the accord had "adequate safeguards against abuse."

Saddam Hussein had read Clinton like a book. He knew that oil prices had risen in 1996. He saw that the U.S. president's desire to keep them down as his reelection approached would make him accept any deal Saddam offered. Which OPEC (Organization of Petroleum Exporting Countries) leader would forget the total disarray into which American politics was thrown by the gas lines, price hikes, and oil shortages of the 1970s, bringing down first Gerald Ford and then Jimmy Carter?

For Clinton, gas prices had a special political piquancy. It was his decision to raise the gasoline tax by a nickel in 1993 that cost him control of Congress in the midterm elections and his increase in car license fees that cost him the Arkansas governorship in 1980 after only two years in office. "Don't mess with their cars" became a political axiom in the Clinton White House.

By first allowing Saddam to sell oil and then by increasing the amount he could export, Clinton was relieving pressure on oil prices. With Republicans embarrassing him by pressing for repeal of his 1993 gas tax hike and pump prices mounting, Clinton doubtless saw the loosening of controls over Iraq as a way out of a tough political problem.

It was one thing to be able to sell oil but quite another to be able to use the money to rearm. To rebuild his military machine and to develop chemical, biological, and nuclear weapons, Saddam had to get rid of the U.N. inspectors. As the *Washington Post* reported, the key protection against misuse of the oil-for-food money was the provision that U.N. "officials monitoring the agreement are given full freedom to travel around Iraq."

Once the inspectors were gone, Saddam correctly reasoned, the restrictions on the use of the oil revenues would become ineffective and he could rearm in peace as he prepared for war.

Saddam started his effort to kick out the inspectors by refusing to allow them access to his dozens of presidential palaces. Then, in November 1997, Saddam announced that he was barring Americans from the U.N. inspection team, denouncing them as "spies." When the U.N. inspectors insisted on keeping U.S. representatives in the group, Iraq barred them all from carrying out their work. In response, the inspectors left Iraq altogether.

The *New York Times* reported that Clinton appeared to respond aggressively by sending "2 aircraft carriers and about 300 warplanes, including the latest F-117A Stealth fighters, plus a score of warships and defense units bristling with Patriot missile batteries, and 18,000 personnel," to the Gulf. In addition, "six B-52 bombers took off from Barksdale Air Force Base in Louisiana for Diego Garcia, the British base in the Indian Ocean. The Pentagon said its 'air expeditionary force,' a special 32-plane combat unit previously announced as being on standby, had been ordered to proceed to the region. . . ."

The old charade—frantic negotiations followed by an apparent concession from Saddam—began again. On November 21, 1997, Saddam seemed to back down and allow U.N. inspectors to return, with Americans among them. In response, Clinton postured, as usual, saying: "Saddam Hussein must comply unconditionally with the will of the international community."

Shrewdly, Francis X. Clines, of the *New York Times*, read between the lines and speculated that the deal "was immediately followed by questions about whether Iraq might have won some secret concessions or understandings through Mr. Hussein's gambit of openly challenging the terms of his defeat in the Persian Gulf war in 1991."

National Security Advisor Sandy Berger defiantly answered, "There is absolutely no understanding. There's no deal. There's no concessions."

Well . . . not so fast. French foreign minister Hubert Vedrine, who criticized Clinton for giving Saddam "the impression that 'there would never be a way out of the tunnel [of sanctions]' even if he got rid of all his weapons programs," noted that "the Americans bent a little" to the demands of Saddam Hussein.

In fact, Berger indicated that an increase in the allowable levels of Iraqi oil sales "might not be opposed by the Administration 'at some point,'" but he hastily added that the subject "never even came up" at the Geneva negotiations.

Six days later, Jim Hoagland pieced more of the story together in the *Washington Post*. Once again, Clinton was being very, very precise in his use of words in order to mislead the American people into believing that he had made no concessions to Iraq in return for the readmission of the inspection team.

Clinton was technically correct—he had made no concessions to

Iraq. He made his concessions to the French and the British, "to allow him to credibly deny making any concessions to . . . Baghdad." In fact, Clinton had agreed to expand the oil sales "if Saddam would rescind his misbehavior over the U.N. inspectors."

Clinton also dropped his earlier insistence that the United States would maintain sanctions as long as Saddam was in power and, Hoagland reported, "raised the threshold for any new U.S. effort to overthrow Saddam to the point of ruling it out," by making clear that he would not attempt to oust Saddam by covert means and that only through a massive American military campaign could the Iraqi dictator be removed from power.

Hoagland's *Post* article explained that the net effect of Clinton's backpedaling was that "under pressure from U.S. allies, Clinton no longer seeks an alternative to Saddam Hussein in Iraq. He is willing to live with a dictator two American presidents have portrayed as a mass murderer days away from creating an arsenal of weapons of mass destruction. On Iraq today, America does not rally the allies, but rallies behind them."

Some contrast with George W. Bush!

Saddam had given up nothing. In return he had gotten his oil sales revenue expanded and pocketed a U.S. guarantee against clandestine efforts to remove him from power. Now he had only to get rid of the pesky inspectors who, the *New York Times* reported, were getting inconveniently close to finding something, having uncovered "stores of deadly nerve agent VX and of botulinus and anthrax toxins."

Back in Iraq, the *Times* reported, the U.N. inspectors walked on eggshells as Iraq insisted that they "should avoid sensitive sites and property belonging to President Saddam Hussein." As the Iraqis put it, the inspectors "should avoid coming near sites which are part of Iraq's sovereignty and national security."

The week after his barring-U.S.-inspectors gambit, Saddam was back with another move. This time, he announced that he would not agree to an extension of the oil-for-food program—in effect, holding his 20 million people hostage—unless the program's restrictions were loosened. Rushing to accommodate him, U.N. secretary-general Kofi Annan indicated that he would suggest raising by 50 percent the amount of oil Iraq could sell, citing reports of starvation among Iraqi children. Saddam now had his daily oil sales up to 1 million barrels per day, a third of his prewar total.

Finally, on November 2, 1998, Saddam Hussein dared to make his big move: He barred U.N. weapons inspectors from continuing their inspections, demanding an end to the trade embargo and a restructuring of the inspection team to reduce the American presence.

Reacting sharply, the U.N. Security Council condemned the dictator's decision and demanded that Iraq let the inspectors resume their work "immediately and unconditionally," insisting that any review of sanctions must come after proof that Iraq had disarmed.

Saddam dug in his heels, sensing the prospect of total victory, and refused to let U.N. inspectors continue to roam Iraq. The inspectors withdrew, and the world waited to see what countermeasures Clinton would order. Would he attack Iraq and demand that inspections resume?

No way. Instead President Clinton and British prime minister Tony Blair tacitly conceded Saddam's ability to oust the inspectors by responding with only four days of intense bombing to protest against his action. American and British troops fired four hundred cruise missiles and two hundred aircraft strikes against Iraq, claiming that it had severely damaged Iraq's ability to produce and repair ballistic missiles, and set back its chemical and biological weapons capabilities.

Cloaking allied impotence in high-flown rhetoric, Blair labeled the new Iraq policy as "containment," stressing that he and the United States were "ready to strike again if Hussein again poses a threat to his neighbors or develops weapons of mass destruction." Blair said that ongoing allied vigilance would keep Hussein "in his cage."

Some cage! Free now to use his oil money to build whatever arms he wanted, Saddam Hussein declared victory. Crowing in a speech to his nation, he said: "You were up to the level that your leadership and brother and comrade Saddam Hussein had hoped you would be at . . . so God rewarded you and delighted your hearts with the crown of victory."

With press and media reports focusing on the intensity of the U.S. and British military strike, Saddam again let Clinton posture while he pocketed his ultimate triumph—the inspectors were gone.

The final nail in the coffin of restrictions on Iraq's oil revenues came in January 1999, when the U.S. ambassador to the United Nations,

Peter Burleigh, agreed to eliminate any limitation on Iraqi oil sales. Nominally, this U.S. concession came to avert a proposal by France, Russia, and China to end the oil embargo altogether. But, as Murkowski put it, "The distinctions between the U.S. plan and the French plan are meaningless. This is the end of the U.N. sanctions regime."

Why Clinton Slept

What accounts for President Clinton's sorry record of weakness in the face of the three-part terrorist threat of al Qaeda, Iraq, and North Korea? Why was Clinton, so aggressive in domestic policy, so reluctant to move to stop terrorism?

At his core, Bill Clinton is a moral relativist. Things are not black and white to the former president; nor do they easily divide into good and evil. Whether facing partisan adversaries or foreign opponents, Clinton could always see the other side's point of view and make allowances for its conduct. Where George W. Bush sees absolutes, Clinton sees complexity.

Shakespeare's Hamlet summed up Clinton's cluttered mind well:

> *Thus conscience does make cowards of us all;*
> *And thus the native hue of resolution*
> *Is sicklied o'er with the pale cast of thought;*
> *And enterprises of great pith and moment,*
> *With this regard, there their currents turn awry,*
> *and lose the name of action.*

Today we call it "paralysis by analysis."

Even after 9/11, Clinton was still seeing the terrorist issue through his opaque lens. As George W. Bush was condemning terrorism as a force that must be obliterated, Clinton provided a window on his more complex and nuanced view of the subject in a speech at Georgetown University on November 7, 2001, barely two months after the attack.

Noting that terrorism "has a very long history, as long as organized combat itself," Clinton reminded his audience of what he labeled *American* terrorism, in an implicit reminder not to see the issue as a simple contrast of good vs. evil. He recited a genealogy of terrorism, from the Crusades through the slave trade and the treatment of Native Americans. Carrying his narrative into the present day, Clinton analogized terrorism to "hate crimes rooted in race, religion, or sexual orientation." The implication was clear: We were not all good, so they could not be all evil.

For all his emphasis on values as president, Clinton was never able to see terrorism as a threat apart from the normal course of international relations. Clinton would not delineate between terrorism and war, nor would he ascribe a motivation as simple as evil to the actions of the other side.

Some who know Clinton well ascribe this lack of dichotomy in his thinking to his relationship with his alcoholic stepfather. Former White House aide Bill Curry has noted that children of alcoholics tend to be lax in reminding their parents of their promises for fear of setting off an alcoholic rage. "I can imagine Bill Clinton's father starting off the day by promising to take him to the movies that evening, only to forget his promise amid his nighttime drinking. Billy would be loathe to remind his dad of the commitment lest he trigger a searing outburst."

In his private dealings, Clinton rarely enforced promises and never saw the transgressions of his staff as grounds for dismissal. Everything was relative. He tolerated an amazing degree of disobedience, disloyalty, conflicts of interest, and untruthfulness in both friend and foe, perhaps accounting for his own tendency to lie and obfuscate. At times it seemed as if the truth had no inherent advantage to recommend itself, but only its relative merit as a practical way to achieve a desired outcome.

While frequently furious at petty slights, Clinton never correlated his anger with policy making. Deliberately, even proudly, Clinton would purge himself of any vestige of rage when he made up his mind to pursue a certain course of action—even when the issues at hand were outrages such as the bombings of the World Trade Center, our African embassies, or American military barracks in Saudi Arabia. Faced with a choice between anger at the perpetrator and empathy for the victim, he always gave emotional priority to the latter.

By contrast, George W. Bush seems to carry a modulated and matured anger into his programmatic deliberations about terrorism and to be unafraid to use it as the basis for making policy. He seems capable of converting the energy of anger into a fuel for decisive action.

Clinton's tendency to moral relativism also handicapped his ability to set proper priorities. Apart from the need to be reelected—and also perhaps to cover up his sexual misconduct—nothing else enjoyed absolute priority in his mind. Terrorism was important, but so were relations with our European allies, civil liberties, budgetary constraints, the price of oil, the starvation of the Iraqi and North Korean peoples, and a host of other considerations, some worthy and others base. Everything was judged in its relation to everything else. Where Bush assigns absolute priority to fighting the war on terror, Clinton could never give anything such unique emphasis.

Nor were Clinton's foreign-policy advisers much better. With the sole exception of Richard Holbrooke, they were an elitist crew determined to keep foreign policy in the hands of professionals. Even such amateurs as former trade lawyer Sandy Berger, Clinton's second-term National Security Advisor, were admitted to the exclusive club of foreign-affairs gurus only if they shed themselves of their tendency to be unduly influenced by the emotions of the common people in the formulation of American foreign policy.

While voters identified terrorism, Iraq, and North Korea as their top foreign-affairs concerns, diplomats like Warren Christopher and Tony Lake were determined to keep things in what they regarded as the proper perspective. They deeply distrusted any excessive zeal in prosecuting Iraq, North Korea, or even al Qaeda as pandering to electoral needs.

Uppermost in their minds was the need to preserve international cohesion in approaching these issues, particularly in our dealings with Iraq. The pro-Iraqi inclinations of the French and the Russians had to be factored in when determining Washington's policy. When Clinton ventured to make his policy of sterner stuff, the threat of a press leak that Clinton was "demagoguing" the issue was enough to hold him in check.

So limited was Clinton's confidence about summoning national resolve for the use of force where there was any real risk of casualties that he always knuckled under in the face of cautious advice from the experts.

My first brush with the arrogance of his foreign-affairs people came as I helped the president prepare his Memorial Day remarks to be delivered at Arlington Cemetery in 1995. I had prepared a draft speech that branded Iraq, Iran, and other nations as international outlaws, linking them to our prior adversaries the Nazis and the Communists. But I was confronted with an angry aide from the Pentagon who told me, bluntly, that if I persisted in pushing my speech draft there would be press leaks that Clinton's political aides were attempting to interfere with the president's remarks on this solemn day of national consecration. Scared off by the threat, Clinton killed my speech draft.

Daunted by a fear that his foreign policy would be perceived as "political," Clinton instructed me never to offer him advice on foreign or military policy matters unless we were alone. Indeed, every week at our strategy meetings in the East Wing, I would bide my time at the end of the meeting until the room was emptied of the others who attended so that I could then sit with Clinton for an hour more discussing international issues. When Sandy Berger, wise to my habits, sought to stay longer to keep me away from Clinton, the president instructed me to pretend to leave the building, then wait downstairs for his all-clear signal so that we could begin our foreign-policy conversation.

When Clinton decided to send ground troops to Bosnia to enforce the peacekeeping deal he had secured after bombing the Serb forces, his foreign-policy advisers insisted that he explain his decision as a move to shore up the NATO alliance. When my polls showed that the public could care less about NATO but was focused instead on preventing more murders and rapes by Bosnian Serb forces, Lake and his aides resisted raising the issue for fear that it would be "pandering" to popular prejudice.

Between the ever-shifting foreign-policy priorities of Tony Lake and Warren Christopher, which blocked decisive action against Iraq and North Korea, and the civil liberties worries of Janet Reno and George Stephanopoulos, which inhibited efforts to stop domestic terror, it seemed as if the entire White House was focused on keeping the president from acting clearly and forcefully to deal with terrorism.

However, none of their efforts would have succeeded but for the fears, worries, and phobias that raged inside Bill Clinton's mind: fear that if he led American troops into a battle with casualties, his own draft record would return to bite him politically; worry that he would alienate his Hispanic constituency if he cracked down on illegal aliens; concern that an increase in the price of oil could spell his political doom; hesitation in the face of European intransigence and worry that his own foreign-policy experts would leak that he was incompetent and too political; willingness to believe he had a deal with North Korea when all he had was a vague and misleading statement of intentions; unwillingness to go to war with Saddam Hussein; trepidation that civil libertarian criticism would undermine his domestic support; and, finally, a morally relativist refusal to see Saddam, al Qaeda, or Kim Jong Il as forces of evil.

These factors, more than any advice from his advisers, paralyzed Bill Clinton's efforts to stem the forces of terror.

By the second half of Clinton's second term, it was too late to focus on terrorism with the intensity the issue required. Disgraced by the Lewinsky scandal, distrusted for lying about his relationship with the intern, hounded by the Republicans during impeachment, Bill Clinton lacked the political and moral authority to stand up to international terror.

Not that he wanted to. As 1998, 1999, and then 2000 brought more and more evidence of an international terrorist conspiracy against America, he became more obsessed with his twin political goals: surviving impeachment and putting his wife in the U.S. Senate.

The White House became a campaign headquarters for Hillary. Bill Clinton had the worst of both worlds—the eroded power of a lame-duck president about to leave office and the timidity of a man focused on the next election. Would an invasion of Afghanistan with ground troops backfire? Was there enough support to pull it off? Would his critics say he was "wagging the dog"—using a war to regain his political footing? Were these risks worth taking as his wife was beginning her political career? No way.

And so Osama bin Laden, Saddam Hussein, and Kim Jong Il lived to fight again another day—against a tougher president.

When Henry Kissinger asked Chinese foreign minister Chou En-Lai what he thought about the French Revolution of 1789, the Communist replied, "It's too soon to tell." We err when we judge a president too quickly after he leaves office. It is only in the hindsight of subsequent events that we understand the wisdom or the folly of his actions.

The success of the containment doctrine in bringing down the Soviet Union gave Harry Truman a vindication that was fifty years in coming.

Vietnam fell, and no other domino keeled over. Thailand, Burma, Malaysia, Singapore, the Philippines—all supposed further casualties of a failure to stop communism in Vietnam—survived our defeat just fine. And when Soviet communism fell fifteen years later, the folly of Johnson's and Nixon's obsessions with Vietnam became apparent to all.

As the 1980s recede into history, Ronald Reagan's efforts to free the economy of government constraints seems wiser and wiser. Japan and Germany, the poster children for planned economies, stagnate, but Reagan's America keeps growing.

Bill Clinton looked a lot better in the White House than he does in the years since. We assumed that he had North Korea under control. He didn't. We let Clinton distract us from Saddam's warlike preparations. We shouldn't have. And we didn't give Osama bin Laden much thought. Big mistake.

In hindsight, Clinton left us naked and unprepared for the perils of terrorism.

For all Clinton's accomplishments (welfare reform, crime reduction, the balanced budget, prosperity, and freer trade), and for all his failures (impeachment, Lewinsky, Paula Jones, the FBI files, Whitewater, and the pardons), it may well be his failure to fight terrorism that will dominate his legacy.

And it should.

FOUR

THE HOLLYWOOD APOLOGISTS

I f there is a war or continued sanctions against Iraq, the blood of Americans and Iraqis alike will be on [American] hands."

"This war is about . . . hegemony, money, power, and oil."

"The war mongers who stole the White House . . . have hijacked a nation's grief and turned it into a perpetual war against any nonwhite country they choose to describe as terrorist."

"There can be no more deaths, no transfusions of blood for oil."

Are these quotes from Saddam Hussein? Osama bin Laden? Are they even the words of French President Jacques Chirac? No. The first three statements are public comments made by actors Sean Penn, Dustin Hoffman, and Woody Harrelson. The fourth is from a Hollywood mega-spectacular, an antiwar statement signed by a parade of stars.

These folks dwell in the rogue state of Hollywood, which shows more evidence each day of breaking off not just physically but mentally from the American mainland and drifting out to la-la land.

And who are these actors and actresses that we should listen to them? On stage, they are human parrots, regurgitating what others have written for them. Off stage, why would we think that they had any wisdom?

Unsuspecting of how they would abuse their prestige, over the years we have let them into our hearts. We sang their songs; tapped our feet to their music; turned the pages of their books, mesmerized by their prose and poetry. We sat enthralled by their movies, thrilled to their action scenes, and wept at their love stories.

But now, those we have supported with our patronage have begun to use that prestige and cultural power against us. They have blamed *us* for 9/11 and criticized our efforts to disarm Iraq, sapping our national will and purpose. They took the entrée we gave them to our souls and used it to sow doubts about our course and trepidation about our journey.

They are the apologists.

Among us still, they drain our energy and blunt our purpose. We need to know who they are, why they oppose us, and where they are wrong, so we may be immune to their blandishments and reject their seductions.

OFF WITH THEIR HEADS!

I believe in the importance of free speech. In my commentary on Fox News and my columns in the *New York Post* and *The Hill,* I live by it. I agree with the remark attributed to Voltaire: "I disapprove of what you say, but I will defend to the death your right to say it." But we still have the right—the duty—to differ with the Hollywood Apologists. Loudly. That's what this chapter is about.

The thing all these well-known celebrities have in common is that they don't know what they are talking about. Most are barely educated and have spent a lifetime learning to mouth somebody else's script, sing a composer's songs, or write novels whose common denominator is that they are, well, fiction.

To suppose that we want to know their opinion of 9/11, or of the war in Iraq, is to assume that celebrity is, in any sense, cerebral. It's not.

Ever since 9/11, there has been a swelling chorus of apologists who either defend those who attacked us or sharply criticize the good vs. evil distinction articulated by President Bush in response to the terrorist assault against us. Some apologists seek to put the terrorism in historical perspective, others to explain its roots, and many justify it by attacking American policy and our friendship with Israel.

Novelist Tom Robbins called the United States of America ". . . an adolescent bully, a pubescent punk who's too big for his britches and too strong for his age." He went on to say that our country was guilty of ". . . mindless, pimple-faced arrogance," and, just as 250,000 Americans were preparing to risk their lives to disarm and depose Saddam Hussein, said, ". . . it might do us a ton of good to have our butts kicked."

Film director Robert Altman was so turned off by America's retaliation after losing 3,100 people that he told a London newspaper, "When I see an American flag flying, it's a joke. There's nothing in America that I would miss at all."

According to leftist chic novelist Gore Vidal, 9/11 was simply a payback for past American actions. "I've listed . . . about four hundred strikes that the government has made on other countries. . . . You keep attacking people for such a long time, one of them is going to get you back."

Somehow, in these commentators' worldview, the United States became the enemy and the American people and our government the perpetrators, not the victims—the bullies, not the ones attacked.

The reality is that never has the United States been so innocent as it was on September 11, 2001. We were peacefully minding our own business when jets ripped into the World Trade Center Towers and the Pentagon. We had done nothing to provoke the attack. *Nothing*.

Yet in the face of all evidence, these apologists dismissed the very idea of our innocence, protesting that we were at fault, that the attack was provoked by our policies, that America had blood on its hands. And when we invaded Iraq, to ensure that other weapons of mass destruction would not be used against us, they claimed we just wanted oil.

Indeed, they seemed to be following one another in a massive rush to emulate the antiwar movement of the 1960s. It was like a Woodstock revival for these former flower children, and *their* children, as they revived all the old slogans and marched as they had in their youth.

This time, however, they were protesting against a just policy forced upon us by outside aggression.

This time, they were wrong.

Apologist Question: Why Do They Hate Us?

Immediately after 9/11, "Why do they hate us?" became the question of choice of the left. Suggesting that 9/11 was somehow something that we had brought upon ourselves, it opened the door to self-flagellation on a mass scale. Over and over again, we heard from prominent liberals that *we* had caused the terrorists to act. Al Qaeda was not simply a band of vicious murderers. No, their actions were apparently a predictable result of American arrogance and superiority.

According to Susan Sontag, author most recently of the novel *In America* and a collection of essays called *Where the Stress Falls*, 9/11 was not a "cowardly attack on 'civilization' or 'liberty' or 'humanity' or 'the free world' but rather an attack on the world's self-proclaimed superpower, undertaken as a consequence of specific American alliances and actions."

Likewise, *60 Minutes* correspondent Andy Rooney suggested that it was our habit of flexing our muscles to show our power that motivated 9/11: "We're puzzled over why so many people in the world hate us, then, next thing you know, we're saying to them, 'Our country is better than your country'—*Yaaaa.* . . . It's as if we're deliberately setting out to make the rest of the world dislike us."

From the left came writer Norman Mailer, who suggested that we caused the terrorists to act: "Everything wrong with America led to the point where the country built that tower of Babel which consequently had to be destroyed. . . . But what if those perpetrators were right and we were not? We have long ago lost the capability to take a calm look at the enormity of our enemy's position."

From the right, the Reverend Jerry Falwell actually told a Christian television audience that the moral tone of America was to blame for the 9/11 attacks: "I really believe that the pagans, and the abortionists, and the feminists, and the gays and the lesbians who are actively trying to make that an alternative lifestyle, the ACLU, People for the American Way, all of them who have tried to secularize America, I point the finger in their face and say, 'You helped this happen.'" Falwell, of course, did not—and could not—explain how people who held these views had helped to cause the death of 3,100 people.

These soppy accusations are as inaccurate and irrelevant as they are masochistic. September 11 should be the occasion for neither self-flagellation nor for evening old scores with domestic political opponents. We were not attacked on 9/11 because of our faults, many though they may be. We were afflicted because of the things we do *right*.

1. The terrorists can't stand our freedom of religion.

They want a world with only one faith and universal conformity of belief. In their universe they are the only believers; all others are infidels. They reject pluralism with a self-righteousness that Western civilization abandoned three hundred years ago. It's worth remembering that religious freedom began not as a right but as a compromise. After two hundred years of religious war, Great Britain finally decided, at the end of the seventeenth century, to allow pluralistic worship rather than suffer the blood baths, first of one side and then of the other, as the changing inheritances of the throne moved her one way and then the other. As a pragmatic step, the Englishmen of the Glorious Revolution voted for religious freedom, removing faith from the national political agenda.

For bin Laden, doctrinal conformity lies at the center of his objectives. In his distorted vision, our infidelity to his dogma is our chief sin.

Not only don't we toe his religious line, but we say so.

2. They hate our freedom of speech.

The same forces that commission hit squads to kill author Salman Rushdie for demeaning the Muslim faith in print want to topple the buildings of the infidel. Ask any of the people of Afghanistan or Iran what happens to free speech when Islamic fundamentalists take over—religious police roam the streets ready to arrest anyone who does or says anything not sanctioned by the clerical authorities.

3. They deplore female equality.

Our treatment of women ranks next in bin Laden's catalogue of our villainy. How odd that so many dedicated feminists should oppose us for standing up to the most antifeminine ideology in the world. We insist on seeing the entire human race as, well, human. We reject the idea that women must be subsets of male desire, covering their faces and hair so as to avoid inciting men to sin. We want women to be doctors and lawyers and presidents—an offense against God, in bin Laden's view.

A statement of support for Afghan women, signed by feminists Gloria Steinem, Eleanor Smeal, Betty Friedan, Catherine Deneuve, and Dorothy Height, issued in June 2000—before 9/11—demanded "the fundamental rights" that have been denied to Afghan women by the Taliban. The statement continued: "The Afghan women reject the false assertions of the Taliban militia that these rights are in contradiction with the religion, culture and traditions of Afghan society and nation. History has demonstrated that supremacist and dictatorial regimes such as the Taliban maintain themselves in power only if the rest of the world remains silent."

Well, we didn't remain silent—we did something about it. And now the Afghan women are free.

4. **They hate that we gave a homeland to the most persecuted people in human history.**

After two thousand years of wandering in foreign nations amid hostile populations, the United States and the rest of the world saw that the Jews deserved a homeland in which they could be safe. Bin Laden's animus is motivated most strongly by a pure anti-Semitism that burns as horrendously as Hitler's.

But apart from misunderstanding why the terrorists hate us, it is maddening that some apologists say that the United States should not be shocked by terrorist acts against us because we have blood on our hands. Some seem to be saying, in effect, "What goes around comes around."

Apologist Line: The United States Has Blood on Its Hands

During his visit to Baghdad, on the eve of war, actor Sean Penn seemed to be doing his best to imitate Jane Fonda, the movie star who famously visited Hanoi at the height of the Vietnam War, posing with antiaircraft batteries that shot down American planes.

What of his statement, quoted at the start of this chapter, that Americans will have "blood on our hands if we war with Iraq"?

Blood on *our* hands? Really? Have we taken a good look at the hands of Saddam Hussein, who:

- killed more than a million of his own people in his wars against the Kurds in the north and the Shiites in the south of his country
- sacrificed hundreds of thousands more in an imperialistic war against Iran
- starved his people by needlessly extending U.N. sanctions simply to maintain his development of weapons of mass destruction
- diverted money from the oil-for-food program to weapons develop-

ment, and diverted enough money to his own pockets that in July 1999 *Forbes* estimated his personal wealth at $6 billion, acquired primarily from oil and smuggling

No, says Mr. Penn, the blood is on our hands. Does he really think we invaded Iraq for sport or for blood lust? We disarmed and deposed Saddam Hussein because we know his weapons threatened the entire civilized world, and we were determined to stop him.

Penn has claimed to be horrified at the concealment of information in the Iraq crisis. He refers in his statement not to cover-ups by Saddam—who shuttled his mobile biological weapons labs around the nation one step ahead of the inspectors—but about *American* cover-ups! In Baghdad, the actor said he "found it 'baffling' that U.S. officials had not revealed more evidence of their suspicions that Iraq has programs to develop weapons of mass destruction."

What is even more incredible than Penn's statements are the coverage they attracted. The *New York Times* devoted a 1,400-word article to the visit of this one movie actor to Baghdad. By definition, it was taking Penn seriously. Why? What was the significance of the actor's statements? Why were they news that is fit to print?

Who cares what Sean Penn thinks? A recent poll showed that two-thirds of those surveyed would rather celebrities "keep their political opinions to themselves," and three-quarters said that the statements of Hollywood types have no effect on their own opinions.

One whose statements do deserve to be taken seriously is former President Bill Clinton. Yet since 9/11 he, too, incredibly, has seemed to imply that there is, somehow, blood on American hands.

Speaking only fifty-eight days after the attacks, Clinton laid out the case that we, too, had been guilty of terrorism in the past. His bill of particulars:

The Crusades: "In the first Crusade, when the Christian soldiers took Jerusalem, they first burned a synagogue with 300 Jews in it, and proceeded to kill every woman and child who was Muslim on the Temple mound."

The truth: The Crusades were a thousand years ago; America wouldn't even come into existence for five hundred more years. How is this relevant, Mr. President?

Slavery: "Here in the United States, we were founded as a nation that practiced slavery and slaves were, quite frequently, killed even though they were innocent."

The truth: Slavery, an abomination we shared with much of the world, came to an end 140 years ago—after millions of Americans shed their blood to end it. As Abraham Lincoln said, we fought ". . . until every drop of blood drawn with the lash" was paid "by another drawn by the sword." Is that not enough to appease and satisfy Bill Clinton?

Native Americans: "This country once looked the other way when significant numbers of Native Americans were dispossessed and killed to get their land or their mineral rights or because they were thought of as less than fully human, and we are still paying the price today."

The truth: The United States did, indeed, wage savage, genocidal war against Native Americans. In a precursor to biological warfare, American troops deliberately gave Indians blankets infected with small-pox as they trekked to their new reservations. But are not these acts of almost two hundred years ago long since expiated?

Lynchings and hate crimes: "Even in the twentieth century in America people were terrorized or killed because of their race. . . . And even today, though we have continued to walk, sometimes to stumble, in the right direction, we still have the occasional hate crime rooted in race, religion, or sexual orientation."

The truth: Bigotry and racism were, indeed, official public policy in most of the South until just forty years ago. Lynching held southern blacks in terror: One count has it that 4,697 fellow Americans were lynched from the late 1800s through the 1920s. But again, these poli-cies were repudiated by the systematic and courageous acts of Ameri-cans, North and South, who purged segregation from the books and still labor to ban it from society. Today our legal system provides elon-gated sentences for the perpetrators of hate crimes.

Sherman's March in the Civil War: "General Sherman practiced a rela-tively mild form of terrorism—he did not kill civilians, but he burned all the farms and then he burned Atlanta, trying to break the spirit of the Confederates. It had nothing whatever to do with winning the Civil War, but it was a story that was told for a hundred years later, and prevented America from coming together as we might otherwise have done."

The truth: Sherman's March did, indeed, shorten the Civil War, saving hundreds of thousands of lives and liberating tens of thousands of slaves along the way.

In likening the terror of 9/11 to these passages in our history, Clinton seemed to be telling us to remember the blood in our own history before we condemn others.

But in each of these cases our nation has fearlessly faced our sins and set about rectifying them, often at a great cost in human life. How do Osama bin Laden and our attackers on 9/11 look upon their acts of terrorism? Do they repent? No, they celebrate their brutal campaign and seek to propagate and continue it throughout the world.

Our national guilt may adhere from history's pages to our consciences, but it is a far cry from those emotions to the celebration of death we see in our terrorist adversaries. And it is an insult to the best Americans—abolitionists, Union soldiers, civil rights activists, and others who fought to better our nation—to dismiss or overlook their history in favor of the worst of their fellow countrymen. It is too bad that Bill Clinton did not articulate that distinction.

Apologist Line: Bush Is Evil and Stupid

The vilification of President George W. Bush by the cerebrally challenged charismatics of Hollywood is quite incredible. Not content with attacking his policies, they belittle his abilities and show arrogant contempt for what they perceive as a lack of intellect.

Among the most vitriolic of the real president's critics is the Hollywood president: Martin Sheen, the president on the television show *The West Wing,* the only government the liberals still control. "George W. Bush," he says, "is like a bad comic working the crowd, a moron, if you'll pardon the expression."

Moron? Really? Mr. Sheen, it should be noted, flunked out of high school and did his higher academic work at acting schools in New York City. The "moron" George W. Bush graduated from Yale University and got his MBA from Harvard, two institutions not known for suffering morons gladly—even if they *are* the sons of alumni and high-ranking politicians.

Sheen went on to say that he felt, in his considered opinion, that "Alcoholics Anonymous and jazz are the only original things of importance" that the United States has exported to the rest of the world. Apparently DNA, the Salk vaccine, the airplane, and the automobile failed to make an impression on Mr. Sheen.

Criticism from a more serious source also challenges Bush's mental outlook. Former South African president Nelson Mandela has called Bush "a president who has no foresight, who cannot think properly, is now wanting to plunge the world into a holocaust."

The idea that Bush wants war, actually desires it, is far-fetched. There can be little doubt that he is as agonized as we all are by the prospect of risking so many lives, all the more so since he will be the one to order it. To suggest that he derives pleasure from his decision and "wants" to plunge "the world into a holocaust" is a moment of undisciplined presumption on the part of Mr. Mandela.

It's easy to dismiss the belittlement of Bush, but some of his celebrity critics go over the top, painting the president as an evil man. The "height of hypocrisy" award must go to David Clennon, star of the CBS TV series *The Agency*. This actor, who makes his living extolling the exploits of the CIA, told Sean Hannity on his February 3, 2002, ABC Radio talk show, "The moral climate within the ruling class in this country is not that different from the moral climate within the ruling class of Hitler's Germany."

When Hannity asked if he was likening Bush to Hitler, Clennon replied: "I'm not comparing Bush to Adolf Hitler—because George Bush, for one thing, is not as smart as Adolf Hitler. And secondly George Bush has much more power than Adolf Hitler ever had. . . . I'm saying that we [Americans] have sunk pretty low and I'm saying that you can look at the moral climate in Germany in 1933. . . . We have to ask ourselves if we found ourselves in Nazi Germany, what would we do?"

If Mr. Clennon rejects the moral climate in Washington, even as he depicts the CIA in a heroic context on television, what would his reaction to Nazism have been? One wonders if he would take a role as a Gestapo agent in a Nazi radio serial?

One thing is sure: If he lived in Nazi Germany, he probably

wouldn't have criticized either the S.S. or the Gestapo on a talk radio show—and lived to tell about it.

When the *New York Times* euphemistically noted that Clennon had spoken out "opposing the war on Sean Hannity's conservative radio show" (a benign characterization of his remarks), the newspaper reported that "he became the object of a campaign to get him fired." The newspaper quoted Mr. Clennon as saying, "I was upset and fearful and angry to be targeted in that way."

Should Clennon be fired? If he worked as a plumber, no way. People in private life should never be fired for their political views or utterances. But when Clennon took a job acting on a prime-time network television show, he left his private privileges behind. Unless CBS wishes to take responsibility for his remarks, he should be dismissed. We all have to take responsibility for what we say, even Mr. Clennon.

But the real question is: Why don't the viewers of *The Agency* deluge CBS with protests and force this man who doesn't know the difference between Bush and Hitler off the air? Why do they remain silent and let him get away with this kind of outrageous comparison?

Right smack in the middle of the first few days of the war in Iraq, Michael Moore, filmmaker and best-selling author, denounced Bush as "a fictitious president" sending us into war for "fictitious reasons" as he picked up his Oscar during the Academy Awards ceremony.

The U.S. Supreme Court and the American people may have accepted George Bush's tenure, but Moore still thinks it's "fictitious." Will Moore also think the weapons of mass destruction we will find in Iraq are fictitious? Mr. Moore's views on the war in Iraq should have been well-known to the Hollywood sponsors of the Academy Awards. The fact that they gave him an Oscar, in the middle of a war he opposed in obnoxious terms, illustrates how far removed the entertainment industry is from the rest of America and how insensitive it can be.

Others find fault with Bush's characterization of the war on terror as a battle of good vs. evil.

In an advertisement in the *Washington Post*, actor Sean Penn, whose best-known credentials remain his performance in *Fast Times at Ridgemont High* and his short-lived role as Madonna's husband, condemned Bush for having "a simplistic and inflammatory view of good and evil."

Harry Belafonte, one heck of a calypso singer, said Bush was "possessed of evil" for preparing for war in Iraq.

Actor George Clooney likened the president to a mobster, telling Charlie Rose, "The government itself is running exactly like the Sopranos." BBC News noted that "[Clooney] said that Bush and his cabinet were behaving like Big Tony and his mobster family."

And what of Barbra Streisand, that famed cheerleader and fundraiser for the left? She claimed to see a lust for totalitarian domination in the president's conduct. She based her charges on a joke Bush told. "Shortly after George W. Bush was elected . . . sort of . . . I saw him on television say: 'If this were a dictatorship, it would be a heck of a lot easier, just so long as I'm the dictator.' And he laughed. You know people never really joke. That was very revealing, a taste of things to come . . . the arrogance of wanting unlimited executive power . . . a government that operates in secret . . . keeping presidential papers secret . . . hiding secret meetings with oil companies. I find George Bush and Dick Cheney frightening. . . . "

One wonders if she was as offended by Hillary Rodham Clinton's efforts to keep the workings of her health-care-reform task force secret in 1994—a project that led to years of litigation before the courts pried open the records.

Among the celebrities who felt themselves qualified to critique American foreign policy, Ms. Streisand deserves special attention for her arrogant ignorance.

Consider the evidence of the limitation of her knowledge:

- In a conversation with the *New York Post,* she referred to former House Minority Leader Richard Gephardt as a senator. When she was criticized for the gaffe, she struck back, saying, in effect, that the *Post* should have covered for her mistake by silently correcting her remarks before they went to press. On her website, she said: "If the *NY Post* does not know that Richard Gephardt is House Minority Leader and that he is not a Senator, how do they know what Barbra Streisand thinks and what she is doing?"

- She used a speech to a Democratic fund-raising dinner to quote a passage she said was from Shakespeare's play *Julius Caesar.* It wasn't. It was an Internet hoax. That the passage, a long soliloquy, was not from Shakespeare would be evident to anyone who had ever read the play.

- She identified Saddam Hussein as dictator of Iran, not Iraq.

- When Matt Drudge mocked her spelling errors in her statements, she replied that she was a spelling-bee champion in school. Not much of a defense: so was Dan Quayle.

- To let us know how important she was, she told *TV Guide* that Vice President Al Gore "called her from Air Force One for advice. I couldn't take the call. I was in the middle of something."

But Barbra isn't alone in her airheaded presumptuousness. No less a moral authority than Natalie Maines, of the country-pop singing group the Dixie Chicks, recently told a live London audience: "Just so you know, we're ashamed the president of the United States is from Texas."

Every one of these aspersions on Bush's sense of democracy is baseless and unfair.

Bush went to Congress to ask for approval of our mission in Afghanistan to retaliate against, and try to capture, bin Laden. He also asked Congress for approval of a war in Iraq and got it. Bill Clinton, by contrast, sought no approval for any of his raids on Iraq, nor for his strikes in Sudan and Afghanistan. Not that he should have, but Bush went the extra mile to consult Congress, a mile that the law didn't require. The provisions of the War Powers Act specifically give the president the full power to commit forces to battle without asking congressional approval for ninety days, after which he is obliged to ask if he still wants to keep the forces in combat. But Bush chose to ask first—twice.

President Bush has specifically sought, and obtained, congressional approval for his efforts to nab terrorists and to hold immigrants for seven days without charging them with a crime. His expansion of wiretap authority not only was approved by Congress, but it represented exactly the same authority Clinton requested after Oklahoma City (which Congress, back then, denied).

Beyond the facts of the matter, though, what makes these tinhorn pundits think they have the moral authority and intellectual stature to deride Bush so personally, calling him evil and likening him to Hitler? From what high perch of ethical or academic achievement do they criticize him? And why do we care what they say?

The media publicize these comments without discussing the stature of those who hurl the accusations. These men and women are in public life. If their utterances are important enough to splash all over the media, then their own intellectual and academic backgrounds deserve examination. What are Martin Sheen's intellectual achievements that he can call Bush a "moron"? What is Barbra Streisand's expertise that allows her to comment on Bush's "arrogance"?

If these people were carpenters, electricians, and bank tellers, nobody would print what they say. So why do we cover them, when they are merely actors and actresses? What makes their opinions worthy of a moment's special consideration?

Apologist Line: Bush Manipulated America's Grief After 9/11 to Support His Personal Agenda

When the planes crashed into the World Trade Center and the Pentagon, Americans were united as we have not been since Pearl Harbor. Dissent was rare, and Americans rallied to support the president. His popularity soared to 90 percent, despite the uncertain mandate with which he won the office. But as President Bush galvanized Americans to invade Afghanistan, to punish and destroy those who attacked us, Hollywood began to criticize him for misleading us.

For example, on February 6, 2003, actor Dustin Hoffman said, "For me as an American, the most painful aspect . . . is that I believe that this administration has taken the events of 9/11 and has manipulated the grief of the country, and I think that's reprehensible."

When a president focuses national anger on achieving the solution to the causes of the pain, he is not manipulating grief but channeling it constructively.

Actress Susan Sarandon charged that Bush has "hijacked our pain, our loss, our fear." This odd and tasteless choice of a word— "hijacked"—has been repeated over and over by the apologists. But how can anyone possibly "hijack" our pain, our fear, our grief? These are internal emotions, unique to the person who feels them. Regrettably, they do not disappear, dissipate, or diminish simply because another also feels them or speaks about them. Whatever else occurs externally, that primitive pain, or fear, remains a leaden weight in our hearts until, as Aeschylus said, "Wisdom comes . . . through the awful grace of God." There is no such thing as "hijacked" emotions. They belong to the holder of the feeling. But the script was written and circulated in Hollywood and parroted loyally by the actors and actresses who work there.

The many actors, writers, musicians et al who have publicly condemned the Bush foreign policy echoed the same theme, saying that "the highest leaders of the land unleashed a spirit of revenge" in response to 9/11 and somehow manipulated the nation's emotions.

What could Bush's motive in "manipulating" our grief have been? The leftists are extremely vague in answering. Some answer the question politically, claiming he was trying to assure his own reelection, after a shaky start, by creating an artificial sense of crisis and by declaring what the actors' and writers' group Not In Our Name has called a "war without limit" against those in the world he labeled terrorists.

Others claim that he sought to justify massive increases in defense spending to feed the appetites of his buddies in the military-industrial complex and repay them for staking his campaign to millions in donations.

Some say that Bush used 9/11 to renew the national focus on removing Saddam Hussein from power in order to pay a family debt, trying to finish the job his father started and for which he was widely criticized after leaving undone. Many even see the issue as a Bush family grudge against Saddam, stemming from the dictator's 1993 effort to assassinate Bush's father during his visit to Kuwait.

Showing a complete lack of understanding of the role of a modern leader in times of crisis, author Susan Sontag has disparagingly described Bush's approach after 9/11 in psychotherapeutic terms, saying: "Those in public office have let us know that they consider their task to be a manipulative one: confidence-building and grief management. Politics, the politics of a democracy—which entails disagreement, which promotes candor—has been replaced by psychotherapy."

But President Bush wasn't manipulating the nation's grief. He was mobilizing it by effectively, accurately, and insightfully identifying the true dimensions of the task at hand, much as Roosevelt did after Pearl Harbor. In one of the boldest presidential decisions of modern times, Bush departed from previous practice when he decided to treat 9/11 not as a crime, but as an act of war. What his critics see as manipulation, I view as courage. Bush understood that the stakes were not just to deter or punish one particular person or group for a specific attack on a given date. He saw that here was an opportunity to ban a tactic of war—terrorism—and eliminate it from the repertoire of rogue nations.

There are precedents.

For example, except for Saddam Hussein, world leaders have largely banned poison gas from warfare. After Germany's massive use of gas as a weapon in World War I, largely against British troops, it fell into disuse. Even Adolf Hitler did not use poison gas on the battlefield in World War II. The Soviet Union did not deploy gas in its Afghanistan War of the 1980s, and the United States did not use it in Vietnam. It was Saddam Hussein who broke the global ban on the use of gas when he shelled the rebellious Kurds and Iranian troops with poison gas in the 1970s and 1980s.

But, generally, gas is eschewed as a weapon in even the most desperate of wars. Why?

Because the nations of the world have indicated such revulsion against its use that they have threatened massive retaliation and the universal ostracism of any nation that dares to break the ban. Indeed, much of the current distrust of Saddam's proclamations of peaceful intentions stems from his decision to use gas in his wars—sometimes against his own people.

Similarly, no nation has used nuclear weapons in conflict since the United States bombed Hiroshima and Nagasaki in Japan, ending World War II. Despite numerous conflicts, no nuclear power has used the bomb, no matter how great the provocation. All understand that once it is used, retaliation becomes a constant threat—a danger too great to treat lightly.

If we can ban poison gas and nuclear weapons, the Bush administration asked, why not terrorism?

Bush realized that no terrorist group could be truly effective without the backing of a nation-state. He understood that, like an AIDS virus, the terrorist group must have a host cell, a nation, in which to implant its destructive ethos and through which to do its dirty work. Had bin Laden not been able to command the support of the Taliban in Afghanistan, he likely could not have launched the 9/11 attacks.

So President Bush wisely decided to focus his attack on eliminating the national backing that terror groups and cells require to operate. He has zeroed in on the nations that are known to encourage, embrace, shelter, and harbor terrorists: the Taliban in Afghanistan, as well as Iraq, Iran, Syria, Yemen, Libya, North Korea, and Sudan. His goal, after 9/11, was to pick them off, one country at a time, to deny terrorists the national base they need for their villainy.

After his initial success in Afghanistan, he has turned to Iraq. Reading the tea leaves, Libya and Sudan have indicated that they want nothing further to do with terrorism. After Iraq, Bush will have Iran, Syria, Yemen, and North Korea on his plate. In all probability, his efforts to curb these nations will also dry up the resources the Palestinians need for their own, vicious, special brand of terror. Bush will also, doubtless, have to act to deter Saudi Arabia from financing international terror as a kind of protection payment to deter revolution against its repressive monarchy at home.

There is nothing manipulative about Bush's policy. He simply understands the global dimensions of the problem and has determined to act broadly and directly to eliminate the threat, rather than to punish one isolated set of actors. That Bush acted this way is not manipulation. Rather, it is evidence of foresight, vision, and determination—qualities to value in a president, not to disparage.

One critic described it best. In the September 16, 2002, issue of *The New Yorker,* Nicholas Lemann observed, "The difference between retaliating against al Qaeda and declaring war on terror is the difference between a response and a doctrine. Beginning with that first speech, Bush has steadily upped the doctrinal ante." Good for him.

Apologist Line: The War on Terror Is Racist

Sometimes you encounter a statement so bizarre that you need to read it two or three times to see if you got it right.

On January 30, 2003, Nelson Mandela wondered aloud why Bush

and British prime minister Tony Blair were "undermining" the United Nations. Was it, he speculated, "because Kofi Annan [the secretary-general] is black?" Mandela noted that "they never did that when the secretary-generals were white." What is there to say in response to such an outlandish comment from the Nobel Prize–winning former leader of South Africa? You shrug your shoulders.

Closer to home, Harry Belafonte had the unmitigated gall to call former general of the army, former national security advisor, and current secretary of state Colin Powell the "house slave" of the Bush administration. How much does an African American need to accomplish, how much does he have to do, before his racial compatriots stop using racist epithets to denigrate his achievements? And who in the world is Harry Belafonte, an entertainer, to call a man who rose from a poor neighborhood of New York City to become a decorated combat commander, chairman of the Joint Chiefs of Staff of the U.S. military, the national security advisor to the president, and the secretary of state a "house slave"?

The more serious racial issue raised by the war on terror is that of ethnic profiling of Arab Americans. The Reverend Al Sharpton, the African American civil rights leader and candidate for president, told a Dartmouth College audience he wanted to "highlight" racial profiling of Arabs as one of his "areas of concern." He said that Bush was using "the current environment to restrict civil liberties." He questioned, in particular, "pulling Arab men in for questioning."

Some apologists have linked post-9/11 targeting of Arab immigrants to what they see as a generic pattern of racism in the United States. Actor Ed Asner, for example, complains that the United States lacks moral authority: "I . . . think that there is a strong streak of racism whenever we engage in foreign adventures. Our whole history in regime change has been of people of different color."

Senator Russ Feingold, who has been working on legislation since 1999 to prohibit racial profiling, notes that "after September 11, the issue has taken on a new context and a new urgency."

And a new justification.

There is no sense in being blind to the fact that Arab immigrants are more likely to hijack airplanes or commit other acts of terrorism than immigrants from other nations or U.S. citizens.

The fact is that the vast bulk of the current crop of terrorism comes from people who are not citizens—who are here, legally and illegally, as immigrants from certain identifiable nations. Indeed, the government has listed Iran, Iraq, Libya, Sudan, and Syria as nations that sponsor terrorism; and Afghanistan, Algeria, Bahrain, Eritrea, Lebanon, Morocco, North Korea, Oman, Pakistan, Qatar, Saudi Arabia, Somalia, Tunisia, United Arab Emirates, and Yemen as countries that harbor them.

Men over sixteen years of age from these nations who are immigrants living in the United States are now required to register with their local INS offices. This is not racial profiling; it is a prudent measure to keep track of guests in this nation who come from countries with a track record that indicates that they may stir up trouble.

On the other hand, though, we must not create an environment in which to be an Arab immigrant—or a U.S. citizen of Arab decent—is to be subject to a host of restrictions and intrusions just because of the color of one's skin.

In some circumstances, profiling is racist and should be banned. In others, it is sensible and necessary.

The distinction relates to *who* is being profiled, *why*, and *where*.

Who is being profiled and searched? The courts have separated the legal rights of U.S. citizens from those of legal immigrants and the rights of illegal immigrants. Profiling that might be illegal for citizens might be acceptable for illegal aliens.

Why are they being profiled and searched? Courts have held that the right to be secure from unreasonable searches and seizures must be balanced against the public purpose behind the action. A simple administrative convenience would not permit such intrusion. But sometimes the public purpose is so compelling as to make the search justified.

For example, clearly the public purpose of stopping drunk driving is sufficient to allow random DWI stops on highways. The courts have ruled that the public purpose of stopping drug use among teenagers is sufficient to permit the requirement of drug testing for students who wish to participate in extracurricular activities at school.

And *where* is the profiling and searching happening? How much privacy can a person legitimately expect at that point? When a person seeks to board an airplane, he has no legitimate expectation of privacy.

He is seeking to use a vulnerable form of public transportation and has to expect to be searched, frisked, and closely examined. If air-security officers seek to single out young men in general and those who are likely of Arab decent in particular for special scrutiny, this is not a decision based on racism. It is based on practical experience.

Obviously, outside a courthouse, it is reasonable to ask those entering to pass through a metal detector. At the same time, of course, it would be unconstitutional to require everyone walking by on the street to pass through one.

But how about stopping a car on the highway because it is driven by a man of a certain ethnicity? Is that racial profiling? What if highway cops have statistics to show that members of certain ethnic groups are more likely to carry drugs in their cars? Should a statistical probability of this sort sanction a search? Of course not.

The difference is the *level* of expectation of privacy. A person driving a car has a lower reason to expect privacy than one quietly sitting at home but more reason than someone seeking to board an airplane. What might be intrusive to demand of a motorist can be acceptable to ask of an airplane passenger.

There's another distinction to be made here, and that concerns the element of choice. If a person sees a metal detector at the entrance to a public building, he has the choice of turning away and not going through it. A passenger seeking to board a plane has the option of deciding not to fly. A student can avoid drug testing by choosing not to participate in extracurricular activities. But it's harder to decide not to drive and impossible to decide not to live in a home. The less the element of choice, the greater the protection against searches and profiling.

And, speaking of racism, how about sexism? Why do so few pay attention to the wonderful work the United States has done in freeing the women of Afghanistan from subjugation? The Taliban regime was, without doubt, the most sexist on the planet, holding women in total bondage and subservience. Their liberation from the strict rules the government imposed led to the joyous scenes of celebration so moving after the end of the war.

Mavis Leno of the Feminist Majority Foundation, and the wife of Jay Leno, is working to get Hollywood celebrities to speak out for the rights of Afghan women. As the *Boston Globe* noted, "The movement to restore human rights to women in Afghanistan has swept Holly-wood."

Now, thanks to the U.S. military and the president who ordered it into action, the women of Afghanistan are on the road to freedom. Does this sound like the work of a reactionary, bigoted nation?

Apologist Line: The War on Terror Is Just an Excuse to Procure Iraqi Oil

George W. Bush's thirst for oil is behind the war on terror: This is a cardinal belief among the apologist community. They each echo the same theme.

Singer Bonnie Raitt is sure that "President Bush is . . . hellbent on protecting access to Iraqi oil and seems willing to risk the lives of thousands . . . to keep gas-guzzling SUVs on the highway."

Linguist and leftist political icon Noam Chomsky, whose misguided views are at least informed by serious thought, writes in his book *Mirror Crack'd:* "The basic reason . . . the U.S. supports corrupt and brutal governments that block democracy and development [is] . , . to protect its interest in Near East oil."

Hundreds of actors, actresses, singers, directors, playwrights, authors, and other cultural icons made the same point in a statement criticizing the war on terror. Signing a statement entitled "Not in Our Name," they declared: "Not in our name will you wage endless war. There can be no more deaths, no more transfusions of blood for oil."

The statement was signed by actors Ossie Davis and Ed Asner, writers Alice Walker, Russell Banks, Barbara Kingsolver, and Grace Paley, playwrights Eve Ensler and Tony Kushner, musicians Laurie Anderson and Mos Def. Other signers included Chomsky, Martin Luther King III, Gloria Steinem, Edward Said, and Rabbi Michael Lerner.

The one thing they have in common is that they don't know what they're talking about.

Barbra Streisand has claimed on her now-infamous website that National Security Advisor Condoleezza Rice "knows the real reason we are invading Iraq now . . . oil"—that it's to create a "distraction" from a failed war on terrorism. Nelson Mandela has also joined this line of criticism, claiming that Bush wants Iraq only for its oil.

Writer Barbara Kingsolver, the author of *The Poisonwood Bible*, said on October 14, 2001 (just a month after 9/11) that "oil gluttony is what got us into this holy war, and it's a deep tar pit."

Oil gluttony? If the United States were motivated by its demand for oil:

- Why has it sided with Israel, which has none, against the Arab states, which have the largest oil reserves in the world?
- Why did the United States favor a total oil embargo on Iraq while Saddam was in power?
- Why did American diplomats resist efforts to permit Iraq to sell oil on the world market?
- When the French and Russians wanted to lift the caps on Iraqi oil sales, why did the United States oppose it?
- Why does the United States unilaterally boycott oil sales from Iran, while its allies all continue to buy their oil?
- Why does the United States prohibit its companies from aiding in Iranian oil exploration and development, while other nations permit and encourage it?

Indeed, if any nations have based their foreign policy on oil, they are France and Russia, who have opposed military action against Saddam Hussein. It is they who are seeking to get oil rights by doing his bidding in international diplomacy.

Let's put the oil issue in perspective. Iraq produces 2 million barrels of oil daily. The total global oil production is 76 million barrels per day. Is the entire war on terror to secure less than 3 percent of the world's oil production?

West Africa, led by Nigeria, is expected to increase its oil production by almost 1 million barrels over the next year or two. Oil exploration is rapidly opening up new sources of petroleum. Indeed, the notorious OPEC has pledged long-term market and price stability, and it has indicated that it will increase its oil production to compensate, should a war knock Iraqi oil supplies off the global market.

Finally, it shouldn't go unnoticed that some of the major terrorist states—Afghanistan and North Korea, for example—produce no oil whatsoever. In Iran, which produces more oil than Iraq, the United

States is not proposing military action and seems content to rely on the normal processes of democratic, domestic change to move the nation away from terrorism.

The United States doesn't need Iraqi oil. That's why we opposed letting them sell any. That can't be the motivation for a war on Iraq. To the Hollywood folks who rant loudly about oil, any conspiracy theory is a good one; any ulterior motive is a plausible explanation. So what if the facts argue otherwise? They're not in the business of facts. Their business is sensation.

Oil is not the issue. Terror is.

Apologist Line: The United States Is Launching a Terrorist Attack to Retaliate for 9/11

Move over, George Bush and Dick Cheney. Shut up, Donald Rumsfeld and Colin Powell. Step aside, Condoleezza Rice. Sheryl Crow has something to say.

Warning darkly of "huge karmic retributions" in the event of U.S. retaliation against terrorism, the Sage of Musicland counseled that "the best way to solve problems is to not have enemies."

Apparently Ms. Crow misled us when she sang "All I want to do is have some fun." She also wants to run foreign policy in her spare time. And almost as absurd as the fact that she offered us this drivel is the idea that our mainstream media covered it. What's nuttier—talking like that or paying attention when others do?

With equal sagacity, actor and noted hemp advocate Woody Harrelson has pronounced solemnly that "the war against terrorism is terrorism."

Is he kidding? Does he really believe that, or is he, as usual, just acting? How is hunting through caves, in search of those who attacked us, a terrorist act? How is trying to maintain homeland security and protect our people terrorism?

Barbara Kingsolver has said that the United States has "answered one terrorist act with another, raining death on the most war-scarred, terrified populace that ever crept to a doorway and looked out." She promises us all that "I am going to have to keep pleading against this madness."

Former senator Adlai E. Stevenson III, son of the Democratic presidential nominee in 1952 and 1956, argues that retaliation against the terrorists of 9/11 might suit the purposes of those who attacked us. "September 11 was not all that different from Sarajevo at the turn of the century. The 19 men armed with box cutters did not expect to bring down all of America. Only America can do that. They expected a reaction. The one they should get is to be treated as criminals, hunted down and brought to justice. Bringing war only confirms complaints that the United States is waging a war against Islam. It can also give terrorists the reaction they seek."

How odd to read the deaths of 3,100 American civilians as comparable to the death of one Austrian prince! Only 3,099 murders wide of the mark, Senator Stevenson. But it's not as if the apologists don't have their own solutions. Actor Richard Gere told a Madison Square Garden audience of rescue workers, firefighters, and police officers that he opposed revenge on the terrorists and that "compassion and understanding" would lead to healing.

Gerda Lerner, emeritus professor of history at the University of Wisconsin, Madison, wrote on October 1, 2001, that "we must be mindful of the danger of becoming terrorists ourselves." While she said that she backed "swift and relentless police action against terrorists," she opposed "terror countered by terror," which only "leads to more terror."

Unlike a lot of the other critics, she had a specific plan for fighting terror. Here it is:

- Pay our long overdue U.N. dues of $2.3 billion.
- Sign U.N. treaties opposing genocide, banning land mines, and banning nuclear and biological weapons
- Strengthen the International Court of Justice and pledge to bring all terrorists we capture under its jurisdiction

Boy, that'll do it all right! Once we catch up on our U.N. dues, those terrorists will disappear!

For the anti–land mine advocates, let me point out that the United States, at great expense, uses land mines that can be controlled remotely and fused or defused at will while still in the ground. Other nations use cheaper, older land mines that blow up whenever a soldier—or a little kid—steps on them. Why should the U.S. military, which can and does disable its mines when hostilities end, ban land mines when they save lives of soldiers and provide a passive defense against aggression. Why don't advocates of a land mine ban focus on mines that cannot be disabled? Because they don't want you to know that the United States uses the more humane kind.

Particularly galling is the tendency of some to see a moral equivalence between the perpetrators and the victims of 9/11.

In the weeks after 9/11, Barbara Kingsolver wrote that "I feel like I'm standing on a playground where the little boys are all screaming at each other: 'He started it!' and throwing rocks that keep taking out another eye, another tooth. I keep looking around for somebody's mother to come on the scene saying 'Boys! Boys! who started it cannot possibly be the issue here, people are getting hurt.'"

One can imagine her parading around the rubble of Ground Zero, making her plea while sullen, grim-faced rescue workers, survivors, and widows and widowers looked on.

If any nation has been victimized, it was the United States on 9/11. We did nothing to cause this attack or to trigger it. We were minding our own business when it came. As we've seen, if anything, our reactions to previous assaults on our sovereignty and lives had been lethargic and unduly passive. To assume that we're just engaged in a circular war of revenge is to miss the entire point of 9/11.

Apologist Line: The U.S. Killed Civilians in Responding to 9/11; We Are the Bully of the World

Speaking of bin Laden after 9/11, author Alice Walker warned that "in a war on Afghanistan, he will either be left alive, while thousands of impoverished, frightened people, most of them women and children and the elderly, are bombed into oblivion around him, or he will be killed."

Medea Benjamin, the 2000 Green Party candidate for senator from California, has said: "We must insist that governments stop taking innocent lives in the name of seeking justice for the loss of other innocent lives."

The United States, in fact, has done everything it possibly can to avoid civilian deaths in both Afghanistan and Iraq. The vast bulk of bombs dropped in both wars have been precision-guided munitions, specifically designed to avoid striking nonmilitary targets. When we consider the tendency of the terrorists to live among civilians, precisely to avoid our attacks, and to use human shields to deter our bombing, the feat is truly extraordinary.

Benjamin also criticized the U.S. action in Afghanistan for dropping "over 20,000 bombs, many of which missed their targets."

While the Pentagon makes no estimate of Afghan civilian casualties, the antiwar group Project for Defense Alternative estimates that a thousand to thirteen hundred were killed in the bombing—a casualty rate of one-quarter of one-hundredth of 1 percent of the Afghan population of 26 million people. Even if we accept the estimates of this admittedly biased source, the U.S. attack on Afghanistan must be rated one of the least destructive of civilian life in the history of modern warfare.

Why was the United States so careful to prevent civilian deaths? Because we are not terrorists. We wanted to *restore* Afghanistan, not destroy it. Those poor people who were killed or wounded by our bombing were the result of our *mistakes,* not of our *intentions.* The terrorists who attacked us planned to kill civilians and timed their attacks during the workday to maximize the number they slaughtered. We did our best not to injure innocent people. This is a deep and abiding difference.

The welcome the Afghan and Iraqi people gave our troops once they had liberated their nations is a testament to the success of the Bush administration's efforts to avoid civilian casualties. The smiles, hugs, flowers, kisses, handshakes, and weeping for joy that greeted our troops as they marched into Kabul and Baghdad reflect not a people shell-shocked by random bombing, but men and women glad to be free and grateful to their liberators: a far cry from the predictions of the doomsayers about civilian deaths.

But the positive record in Afghanistan in avoiding civilian deaths hasn't deterred know-nothing Hollywood types from piously declaring, in their "Not in Our Name" petition: "Not in our name will you invade countries, bomb civilians, kill more children, letting history take its course over the graves of the nameless."

Singer Bonnie Raitt has said, "More than a decade of U.S. efforts to undermine the regime of Saddam Hussein has produced utter misery for Iraq's 23 million people. A renewed military campaign by the United States against Hussein would wreak further havoc and devastation on Iraq's population."

Ms. Raitt does not seem to understand that the misery of the Iraqi people is due not to U.S. military action or even economic sanctions, but to the determination of Saddam Hussein to steal his people's money to rearm. Throughout the 1990s, as we've seen, Saddam hoodwinked the United Nations into allowing Iraq to sell progressively more oil as part of its humanitarian oil-for-food program. But Saddam Hussein did not use the money to feed his people. Instead he spent all he could to develop his military capability, and in particular to build nuclear, biological, and chemical weapons. The misery of the Iraqi people is not due to American action but to the work of their nation's own leader.

Perhaps Ms. Raitt's policy briefings did not leave her sufficiently informed about Saddam's extermination of 180,000 young Kurds who, he felt, might endanger his regime in the future—or the more than 1 million people killed in the war he initiated with Iran.

The ubiquitous Woody Harrelson told the British newspaper the *Guardian* a touching story of an encounter with Bill Clinton during a visit to the White House. Harrelson said Clinton told him, "Everybody is telling me to bomb" Saddam (after he threw out the U.N. inspectors). "All the military are saying, 'You gotta bomb him,'" Harrelson said Clinton told him. "But if even one innocent person died, I couldn't bear it," Clinton continued. Harrelson relates how "I looked in his eyes and I believed him." (So much for his astuteness). Harrelson adds: "Little did I know that he was blocking humanitarian aid at the time, allowing the deaths of thousands of innocent people."

The problem was, Woody, Clinton wasn't blocking humanitarian aid. The United States had just agreed to let Iraq sell about all the oil it could. The reason we wanted the inspectors to stay in Iraq was to be sure the money went to *feeding kids*, not to producing weapons. Harrelson said it was "a cowardly act" for the United States to drop "cluster bombs from 30,000 feet on a city." Cluster bombs? Once again, Harrelson showed his ignorance. The United States has not used cluster bombs, in the sense Harrelson spoke about, since Vietnam. In fact, we have specific prohibitions against their use by recipients of our military aid—despite the fact that this policy has become a point of some friction with Israel. We actually use the *opposite* of cluster bombs—precision-guided bombs, through which we do our best, at a huge cost for each bomb, *not* to hit civilians or to kill randomly. The cluster bombs that we used in Iraq sprayed precision-guided bombs over the battlefield. Each were equipped with special sensors to find tanks. If there were none, the bombs fell harmlessly to the ground and did not explode.

Perhaps the all-time award for irresponsibility, though, goes to actor George Clooney, who said, on Charlie Rose's PBS show: "We're picking on people we can beat, you know, and we're saying 'okay, we'll go get them.'" Has Clooney noticed that our enemies have something else in common—they sponsor terrorism and are actively developing weapons intended to destroy us? Another actor who is reading his script without knowing what he is talking about.

Sometimes the blame-American crowd take their opposition to American military action in Iraq awfully far. After calling the United States an "adolescent bully," novelist Tom Robbins said, "Quite probably the worst thing about the inevitable and totally unjustifiable war with Iraq is that there's no chance the U.S. might lose it."

Robbins goes on to get himself into even more trouble by saying, "It might do us a ton of good to have our butts kicked. Unfortunately, like most of the targets we pick on, Iraq is much too weak to give us the thrashing our continuously overbearing behavior deserves, while Saddam is even less deserving of victory than Bush."

What a ringing endorsement of the president of the United States!

Then Robbins offers his boilerplate denial: "Don't get me wrong—I don't want American soldiers killed. But I don't want Iraqis killed, either. I'm just not one of those people who believes that American lives are more valuable than the lives of others." But how does he propose to have "our butts kicked" without tens of thousands of American dead?

Apologist Line: It's Israel, Stupid

Osama bin Laden frequently cites U.S. support for Israel as a key justification for his attack on 9/11. His linkage of terrorism against the U.S. and the Palestinian cause finds wide echo throughout the Arab world. It is hard to forget the scenes of West Bank jubilation among Palestinians who took to the streets to celebrate the atrocities of 9/11.

More disturbing is the implication among some Americans that the United States should be more even-handed in dealing with Israel and its Palestinian assailants. Former U.S. senator Adlai E. Stevenson III has said that "the Israeli-Palestinian conflict is a good place to begin" in understanding the cause for our unpopularity. "The United States loses credibility," he explained, "when perceived as supporting terror in one part of the Mideast while professing to fight it elsewhere."

The "Not In Our Name" (NION) statement, which is critical of American Middle Eastern policy in general, and the proposed war in Iraq in particular, takes special aim at Israel. It noted the "brutal repercussions" of U.S. policy felt "from the Philippines to Palestine, where Israeli tanks and bulldozers have left a terrible trail of death and destruction."

Saying that they "draw inspiration from the Israeli reservists who . . . refuse to serve in the occupation of the West Bank and Gaza," the signers declare that they will "repudiate any inference that [these acts of war] are being waged in our name or for our welfare. We extend a hand to those around the world suffering from these policies; we will show our solidarity in word and deed."

In rebuking Israeli "tanks and bulldozers" for the "terrible trail of death and destruction" they have left in their wake, the "NION" signers make no reference at all to the Palestinian suicide/homicide bombers whose violence precipitated the Israeli retaliation. More than seven hundred Israelis have died in suicide/homicide bombings. Proportionately, it would be as if seventy thousand Americans had met their death in such fashion—more than we lost in Vietnam. Doesn't Israel have the right to retaliate?

What of the blown-up school buses? How about the incinerated restaurants filled with diners? Who is to be called to account for the community centers, filled with elderly, set afire?

Is there anyone among us who thinks that Israel would have invaded the West Bank with tanks and bulldozers and leveled Palestinian villages and camps had they not been provoked by these suicide/homicide bombings? Does anyone think that Israel retaliates just for fun? Obviously, the Israelis are motivated only by a desire to stop future attacks, by disabling the would-be bombers before they bring their cargo of death, strapped to their bodies, over the Israeli border.

When Stevenson calls American policy in the Middle East "supporting terror," he confuses the violence of the terrorist with that of the counterterrorist. Is it wrong to attack the locations where bombs are made and terrorists trained before they can be unleashed? Does Israel not have the same basic national right of self-defense that we do?

Where does political opposition to the U.S. support for Israel end and anti-Semitism begin? Anti-Semitism has always advanced in disguise; rarely have historical oppressors attacked Jews in public when they could find a pretense to do so under the cover of legitimacy. History is filled with attacks on "Jewish money lenders," "the International Jewish conspiracy," and "Jewish communists and leftists." Now, chic anti-Israeli opinion increasingly summons memories of these earlier bouts of anti-Semitic global opinion.

There's no need to split hairs in defining anti-Semitism. When these luminous authors, actors, playwrights, musicians, and academics speak up for the Palestinians, they are endorsing mobs whose favorite slogans are "kill the Jews" and "death to the infidel." In supporting an overtly anti-Semitic effort against the Jewish community, they are doing the work of anti-Semites throughout the ages. One would have thought that chic, fashionable anti-Semitism had not survived the death camps of World War II—but apparently that was too much to hope for.

Let us not forget how Israel came to be in the first place. After the extermination of 6 million Jews in Hitler's death camps, the world community came to realize that Jews were a people without a country, subject to the prejudices and bigotry of others from the Inquisition down through the concentration camps.

Jews became stateless because of the aggressions of the Roman Empire. But they became hated largely through the deliberate policies of the Roman Catholic Church and other Christian denominations, which branded Jews as "Christ killers." As James Carroll documents in his book *Constantine's Sword*, Jews have suffered as a result of a systematic effort to scapegoat them for things that neither they, nor their ancestors, did. Modern papal efforts to obliterate the scar of these past accusations have done little to lift the curse of anti-Semitism from the Jewish people.

The consensus grew after World War II among international political leaders—Adlai E. Stevenson II (III's father) among them—that Jews had to be given a state. The United Nations then voted to partition the British territory in the Middle East into Israel and Jordan, creating two new states with new boundaries.

How did Israel come to expand? How did it go from the small nation envisioned by the United Nations to the more expansive boundaries of the modern state? It did it the old-fashioned way—through war. But these were wars initiated not by Israel but by Arabs, including one notorious surprise attack on Yom Kippur, the highest of holy days in the Jewish religion.

In four wars, Israel was attacked by a coalition of all the Arab nations. Outnumbered hundreds to one, the tiny Jewish nation fought for its life with a tenacity and determination catalyzed by its collective memory of parents going helplessly into the gas chambers of Nazi Germany.

In 1967, the third of these wars started and planned by the Arabs resulted in a sweeping Israeli victory. Its troops occupied, as they had in both previous wars, the West Bank of the Jordan River and the Gaza Strip, a tiny protuberance of land jutting into Israel from the Sinai Desert. In the two previous conflicts, Israel had given back the conquered land; each time, though, the Arabs had reoccupied it and used it as a springboard for another invasion. In 1967, they hung on to it and protected it against yet another Arab invasion in 1973.

So why do liberals and intellectuals stand up for the Palestinians and show so little concern for the Jewish people, who have suffered far worse and for far longer? I believe anti-Semitism is a big part of the answer. Any objective assessment of the Middle East has to lead to a few conclusions:

- It was a good idea to found Israel.
- Israel wants to be left in peace.
- It is the Palestinians who are initiating the violence that plagues the region.

The United States deserves not blame but unstinting praise for helping to found and support Israel, to protect the most persecuted people in the history of mankind.

Apologist Line: Poverty Is the Breeding Ground for Terrorism

If the extreme right robotically blames any ill on immorality, the left looks for explanations in global poverty. Adlai Stevenson III's knee jerked after the stimulus of 9/11 when he wrote in the *New York Times:* "Whether made by al Qaeda or Saddam Hussein, today's threats require a multidimensional response, including efforts to address the widening gap between the haves and the have nots, the horrible conditions in which most people around the world struggle to survive."

Feminist writer Gerda Lerner is most explicit in drawing a connection between the 9/11 attacks and the poverty of the Palestinian refugee camps. She says, "Unless we attack the causes of worldwide terrorism, our capture of a few of its leaders will be an empty victory. One of the major breeding grounds for terrorism has been the existence of refugee camps in which whole populations linger for one or more generations, without outlet, without education, and without hope."

Her language is elegant, but can one really say that Osama bin Laden is "without education" or a member of the world's poor? The Arabs who toppled the Twin Towers were mostly Saudis, many from upper-middle-class families, not the poor who "breed" violence in refugee camps.

The fact is, bin Laden is reputed to be one of the richest men in the world, with assets once valued in the hundreds of millions. In October 2001, *Time* reported that bin Laden is "the son of a billionaire Saudi construction magnate, he has an estimated net worth of hundreds of millions of dollars, including real estate in Paris, London and the Cote d'Azur, and as much as $150 million in stock."

This victim of poverty "runs a portfolio of legitimate businesses across North Africa and the Middle East. Companies in sectors ranging from shipping to agriculture to investment banking throw off profits while also providing cover for al Qaeda's movement of soldiers and procurement of weapons and chemicals."

This is hardly the dossier of a man embittered by his own poverty. Poverty had nothing to do with bin Laden's misanthropic decision to kill as many Westerners as he could.

Ms. Lerner also ignores the vast amount of money spent by the global community on aiding the Palestinian refugee camps. In 2002, the U.N. Relief and Works Agency for Palestine Refugees in the Near East (UNRWA) will spend $296 million alleviating the poverty of which Ms. Lerner complains. With this aid lavished on an estimated 4 million refugees, the United Nations is doing a great deal to mitigate the poverty of the camps.

If poverty persists, as it does, it is not due to the lack of generosity of the United States and the Western world, but to the corruption of Yasser Arafat's regime, which uses the funds to pay for its own luxurious lifestyle at the expense of its people. Indeed, *Forbes* recently listed Arafat as another of the richest men in the world, estimating his personal fortune at $300 million, every penny probably taken from the mouths of his people through corruption.

The international terrorists of the Middle East are not Robin Hoods but robbing hoodlums.

Apologist Line: The War on Terror Is Leading to a Trampling of Our Civil Liberties

The *New York Times* wasn't alone in its alarmist concern about civil liberties after 9/11. A noisome rally of protest on the subject came from the Hollywood apologists, who charged that our attempts to protect America against terrorists was about to jeopardize the civil liberties of thousands of innocent citizens.

Barbra Streisand was moved to say: "You can't defend America by attacking the very rights and liberties that we're fighting for." At a Democratic fund-raiser, she said: "I find the erosion of our civil liberties in the guise of homeland security frightening." Streisand and others were worried that those suspected of terror would not be given the same constitutional protections as others.

The concern was reiterated by the Not in Our Name coalition, who protested that Bush had created two classes of people: "those to whom the basic rights of the U.S. legal system are at least promised, and those who now seem to have no rights at all." They proclaimed: "Not in our name will you erode the very freedoms you have claimed to fight for."

Many of the concerns derive from a shift in the emphasis of government investigation, from *prosecution* of particular suspects for specific crimes to a generic attempt to get *information* with which to protect against another terrorist attack like 9/11.

In the Clinton administration, the emphasis was on prosecution. Now, after 9/11, it must be on intelligence gathering if we are to prevent another attack. As Attorney General John Ashcroft has said: "We haven't forsaken [prosecution] as an objective, but our priority has to be to prevent, to curtail, to disrupt, to interrupt, to keep from happening again the kind of event that could take another [3,100] lives."

When the government is prosecuting suspects, it must be closely bound by the protection afforded defendants by the search-and-seizure prohibitions in the Fourth Amendment and the exclusionary rule that keeps illegally obtained evidence out of court. However, when the federal agencies are not prosecuting individuals but gathering intelligence to stop attacks, it must be free of Fourth Amendment constraints. After all, the Constitution is designed to protect individuals from arbitrary government prosecutions. But when nobody is in the dock and the effort is to prevent attacks, a different standard must prevail.

The Foreign Intelligence Surveillance Act (FISA), according to Evan P. Schultz in the *Legal Times*, provides that "government officials, so long as their primary purpose is intelligence gathering, can obtain warrants from a special court without any adversarial arguments or briefs, in secret hearings, for broad searches and seizures of communications and objects." Schultz bemoans the civil liberties implications of FISA, saying that "the government does not necessarily care about putting [terrorist suspects] on trial or in prison. So much for the exclusionary rule."

Wait a minute here! The goal of the Constitution is to protect the individual, not to make life harder for the government. If the United States uses illegally obtained evidence to put a man or a woman in jail, that should be a violation of the Constitution, and the evidence should be tossed out under the exclusionary rule. But if we use those same tactics to get information and intelligence to foil another terrorist attack, and it has nothing to do with locking anybody up, what's the problem? The Constitution protects the rights of the individual, not of his cause!

The new rules do nothing at all to weaken the evidentiary standard in criminal trials. The evidence that the government gains in its intelligence gathering may not be admissible in court in a prosecution. For that, the government must still collect its evidence the same way it always has.

The sole exception is the provision in the new Patriot Act that would, in the words of Wisconsin's Democratic Senator Russ Feingold, allow prosecutors to "use . . . information obtained by foreign law enforcement agencies in wiretaps that would be illegal in this country."

So what? If the evidence is gathered on foreign soil by foreign police, why not allow its use in an American court? We exclude evidence that is illegally gathered to punish American cops for unconstitutional actions, to ensure that they have an incentive to obey the Constitution. But with foreign police, where is the public purpose in excluding the evidence? Does any American citizen have the right to expect that his American constitutional rights will be protected by foreign police on foreign soil? If he does, he's a fool.

Feingold is also particularly incensed that under the administration's antiterror bill, searches can be conducted without serving a warrant. Understand, agents still have to *get* a warrant. They merely aren't required to *serve* it on a suspect, if the government can show it has "reasonable cause to believe" that serving the warrant and providing notice of the search may "seriously jeopardize an investigation." The senator said the provision was "a significant infringement on personal liberty."

Really? If a law-enforcement agent must tell a judge why he needs a warrant and the magistrate issues one, what purpose is served by informing the terrorist suspect that his home has been searched while he was away? The only thing that serving the warrant would do is alert the suspect that he is under investigation and give him and his confederates a chance to destroy evidence.

The senator's finickiness about serving warrants might be better placed in trying to help agents of federal investigative agencies give warning to Americans who might become victims of terror.

Feingold was also outraged by a provision that broadened the criminal forfeiture laws "to permit, prior to conviction, the freezing of assets" of the defendant even if they were unrelated to an alleged crime. We do just this in drug cases. Is terror less important in the senator's view?

Feingold also said he was "troubled" by "a provision that permits the government, under FISA, to compel the production of records from any business regarding any person if that information is sought in connection with an investigation of terrorism or espionage." This means, he claims, that the "government can apparently go on a fishing expedition."

But guess what? I *want* our government to do all the fishing it needs, if it means hauling up evidence of an impending terror plot. It isn't easy to protect a nation of 285 million people. It's impossible to do it without compromising at least the fringes of our privacy. And, in every legal analysis of privacy rights, a balancing test is always invoked. In any given instance, is the national interest more important than the individual's right to privacy? In other words, what are the stakes?

Some people don't agree with me. Many have complained that the government might be able to see what library books you borrowed or what websites you logged on to. I say, "So what?" I trust our government a lot more than I do the terrorists. It all boils down to whether you believe your government has it in for you and will use any excuse to pry out your most personal information to zap you. I don't. I want my government to have all these powers to fight terror, and I trust it to use them responsibly.

Feingold objected to giving "the attorney general extraordinary powers to detain immigrants [for seven days] . . . on [the] mere suspicion that the person is engaged in terrorism." Feingold criticized the bill for its "deep unfairness" in using "an immigration status violation, such as overstaying a visa" to hold people suspected of terror.

We owe no duty to people who are here illegally. There is no "deep unfairness" in holding someone who is here in violation of the law, by overstaying their visa for example, if the attorney general thinks some important public purpose can be achieved by keeping him in custody. If the illegal immigrant doesn't want to be subject to detention, then he can try obeying our laws and leave when his visa expires.

Much ado has been made of the attorney general's policy of eavesdropping on conversations between terror suspects and their attorneys. Ashcroft has used this power against only 16 of the 158,000 federal inmates—because, he says, "we suspect that these communications are facilitating acts of terrorism." Besides, "each prisoner has been told in advance his conversations will be monitored. None of the information that is protected by attorney-client privilege may be used for prosecution. Information will only be used to stop . . . terrorist acts and save American lives."

Ashcroft's insistence on monitoring attorney conversations with their terrorist-suspect clients was vindicated on April 10, 2002, when an attorney from Brooklyn, New York, Lynne Stewart, was arrested for allegedly passing orders from her client, Egyptian cleric Sheik Omar Abdel Rahman, to terrorist groups beyond his prison walls. Rahman is serving a life sentence for his role in the aborted plot to blow up the United Nations, the Lincoln and Holland Tunnels, and the George Washington Bridge in New York City.

The indictment charges that Stewart ferried letters from an Islamic terrorist group regarding a possible resumption of "military operations" and that she passed a message from Rahman urging "the Muslim nation to fight against the Jews and to kill them wherever they are."

According to CNN, the evidence for the indictment came from "hundreds of intercepted conversations . . . obtained by eavesdropping inside prison and from wiretaps . . . [on] phones and computers in their [the terrorists'] homes."

Ashcroft also sent FBI agents to interview immigrants throughout the United States to gather information about terrorism. He explains: "We have asked a very limited number of individuals—visitors to our country holding passports from countries with active al Qaeda operations—to speak voluntarily to law enforcement. We are forcing them to do nothing. We are merely asking them to do the right thing: to willingly disclose information they may have of terrorist threats to the lives and safety of all people in the United States.

What, precisely, is wrong with asking the FBI to investigate?

A particular focus of the civil libertarians has been the six hundred Taliban and al Qaeda fighters (from forty different countries) now being held at the Guantanamo Bay U.S. military base in Cuba. These are not just foreign nationals. They are terrorists from other countries brought here after being taken prisoner by American forces. Most either fired on U.S. troops or were arrested after being tracked down by the FBI. The government refuses to call these captives prisoners of war, since they do not represent any national military structure. Federal district court judge Colleen Kollar-Kotelly ruled that the men cannot be tried before a

U.S. court, since our legal system has no jurisdiction over Guantanamo. Amnesty International accuses the United States of keeping them in a "legal black hole."

Since none of these men are U.S. citizens, and none were on American soil when they were captured, there is no reason they should be afforded the protection normally given American citizens before the criminal justice process. Our goal is not necessarily to prosecute them but to use the incarceration—and the hope of release—to get information to prevent other terrorist activities.

Colonel John Perrone, a former joint detention commander at the camp, points out: "There is no torture at Guantanamo Bay, but without the carrot of release it is not easy to get information from the men. Sometimes it is a very tedious process and it takes a great deal of time for information to turn out to be fruitful." BBC News reported that U.S. authorities say that intelligence gleaned from repeated questionings has helped prevent terrorist attacks around the world. "There are reports that an al Qaeda plot to blow up warships in the Straits of Gibraltar was foiled after information was obtained from a detainee in Guantanamo Bay."

Why should the United States have to release or charge these captives? Why should they have the protection of our legal system? They aren't citizens, they weren't on U.S. soil, and their only connection to the United States was that they fired on our troops. To prosecute them in open court would entail giving up intelligence sources that could be important to preventing future attacks. In past wars, prisoners were held until the war ended or until their release served U.S. purposes. That sounds like a good idea this time, too.

After 9/11, a massive, worldwide dragnet led to the arrest of twenty-four hundred people in ninety countries on suspicion of terrorism. In the United States, about twelve hundred Middle Eastern or South Asian men were secretly arrested and held right after September 11. Most were deported. Civil liberties groups objected strongly, but there is no evidence that the government abused its power. Those whom it could deport, it did. Nothing wrong with that. Those who belonged here were released shortly.

When civil liberties need protecting, our system is doing a good job of making sure they are. The Pentagon probably overreached with its proposal for a Total Information Awareness program to check for suspicious patterns in Internet and other communications. According to the *New York Times* of February 12, 2003, the system "would enable a team of intelligence analysts to gather and view information from databases, pursue links between individuals and groups, respond to automatic alerts, and share information, all from their individual computers. It could link such different electronic sources as video feeds from airport surveillance cameras, credit card transactions, airline reservations, and records of telephone calls. The data would be filtered through software that would constantly seek suspicious patterns."

As a result of congressional oversight, the Pentagon was denied the ability to use the system where American citizens are concerned. They could, however, use it to keep up with the doings of noncitizens, an appropriate distinction. If you're looking for reassurance that the government will keep a check on its own antiterror measures when it comes to civil liberties, there's no clearer evidence than its self-imposed curb on the Total Information Awareness program.

Apologist Line: The Hijackers Were Not Cowards. But the United States Is Cowardly for Its Stand-Off Bombing and Other Military Tactics

Sometimes public policy debate gets downright silly—as it did in the early days after 9/11, when the apologists objected to the characterization of the hijackers as cowards.

Soon after the attacks, *Politically Incorrect* talk show host Bill Maher compared the courage of the hijackers with that of America's military leaders: "We [the United States] have been the cowards lobbing cruise missiles from two thousand miles away. That's cowardly. Staying in the airplane when it hits the building—say what you want about it, it's not cowardly."

The firestorm that followed cost him his show.

Author Susan Sontag took the same ridiculous theme, saying, "If the word 'cowardly' is to be used, it might be more aptly applied to those who kill from beyond the range of retaliation, high in the sky, than to those willing to die themselves in order to kill others. In the matter of courage . . . whatever may be said of the perpetrators of Tuesday's [9/11] slaughter, they were not cowards."

This macho posturing is beneath even the dignity of the apologists!

There is no bad name, no curse, no evil connotation in the English vocabulary I would not happily apply to the nineteen hijackers of 9/11. To rub salt into the wounds of so many who were grieving by, in effect, praising the courage of the hijackers is insensitive, cruel, vicious, and contemptuous of their suffering.

But to perpetrate the fiction, the canard, the fraud that the American military is "cowardly" because it seeks to minimize casualties is not only wrong, it is downright harmful. Our servicemen and -women have no need to prove to anyone their courage, their willingness to face danger. To say that efforts to protect their lives are cowardly is unconscionable.

Thank goodness the American military took to heart the key lesson of Vietnam: We must find new ways to win wars without killing tens of thousands of Americans. As a nation, as a democracy, we are simply not willing to part with fifty-eight thousand young men and women again. We will fight for justice and peace abroad, but we object to an exorbitant price in American blood.

By using remotely piloted aircraft, as opposed to manned bombers, we avoided losing any prisoners to the enemy in the Afghan War. The issue of POWs, so distracting in Vietnam that the war was prolonged for years while the two sides haggled over their fate, never arose in Afghanistan. The number of American soldiers captured was also held to a minimum in Iraq because our military had figured out how to use unmanned drone aircraft for many of its most dangerous missions. This is not cowardice—it is good policy.

Similarly, stand-off bombing, cruise missiles, night fighting, extensive body armor, helicopters ready for evacuation, and a host of other tactical improvements since Vietnam held American combat deaths in Afghanistan to forty men and women.

Indeed, the total of American combat deaths in all U.S. military actions since 1980—Panama, Haiti, Grenada, Libya, Bosnia, Sudan, Kosovo, Afghanistan, the Philippines, and operations Desert Storm and Iraqi Freedom in Iraq—is fewer than one thousand soldiers. Thank the Lord. And thank our military planners who sought a way to protect freedom without sacrificing too many of our best men and women in the process.

But perhaps the single most offensive comment of all the statements made by the 9/11 apologists—a distinction indeed—was by filmmaker and best-selling author Michael Moore, who called the passengers—the victims—on the hijacked airplanes "scaredy cats." In a bizarre attempt at humor, Moore claimed that this was because they were white. "If the passengers had included black men," he said, "those killers, with their puny bodies and unimpressive small knives, would have been crushed by the dudes, who, as we all know, take no disrespect from anybody."

If he were serious, this comment would be an outrage. The fact that he was trying to be funny makes it even more so. Are not the bodies of those who gave their lives to save the Capitol or the White House above such desecration?

Talk about a stupid white man!

Apologist Line: We Are Ignoring North Korea Because of Bush's Obsession with Iraq

"Secretary of State" Dustin Hoffman has demanded to know why we are focusing first on Iraq and giving insufficient attention to North Korea, another rogue state seeking nuclear weapons. The "secretary" said: "If they [the Bush administration] are saying it's about the fact they [Iraq] have biological weapons and might have nuclear weapons and that gives us the liberty to preempt and strike because we think they might hit us, then what prevents Pakistan from attacking India, what prevents India from attacking Pakistan, what prevents us from going into North Korea?"

Hoffman seems to misunderstand his own point: Pakistan, India, and, most likely, North Korea, already *have* nuclear weapons. That's why they haven't attacked one another; that's why we're not approaching the challenge of dealing with them the same way we are Iraq. One can have considerably more freedom of action with horrendous dictators who do not have the bomb than one can when they get it. Pyongyang may have the bomb; Iraq doesn't. With rockets and artillery guns able to hit Tokyo and Seoul and, perhaps, nuclear warheads with which to equip them, North Korea is a menace we must approach gingerly.

That was one of the best reasons to disarm Saddam before he got the bomb. But the apologists aren't concerned with history or with facts. What they're concerned with are the openings on the talk shows instead—openings that allow them to take what sound like eloquent and noble stances against imaginary evils. What they fail to realize is how these stances undercut the efforts of the very nation that is fighting to defend them, their families—yes, even their careers.

The U.S. government isn't perfect; its actions aren't always above error or suspicion. It is a government led by flawed human beings who make mistakes. But its intentions are good, its people are virtuous—and its goals, right now, are unmistakably urgent. It is worth trusting the Bush-Cheney-Rice-Powell-Rumsfeld-Ridge-Ashcroft administration to make the right decisions. If they put a foot wrong somewhere along the way, they will have done so for the best of motives.

One can disagree with this administration on a host of domestic issues. I sharply dissent from its policies on the environment, health care, Social Security, Medicare, campaign-finance reform, tort reform, and a wide range of other issues. But in the war on terror, the leaders of our government know what they are doing. And what they're doing is basically right and deserves our full, though not unquestioning, support.

The real question posed by the parade of celebrities leading the opposition to the war on terror is not their right to speak; nor is it even what they have to say. It is why we cover them—why we pay attention in the first place. Somewhere along the way—is it because they all come into our homes through the same TV screens—we seem to have developed the impression that the opinions of celebrities and those of credentialed experts were created equal. Why should the media report what Bonnie Raitt or Sheryl Crow says about the war on terror? What can it possibly matter what Barbara Streisand thinks? Do we listen to what Henry Kissinger thinks about the New York Yankees this season? What Hillary Rodham Clinton has to say about football plays?

In this regard, every link in the media food chain—the celebrities who serve up half-baked "political" commentary, the media who package and purvey it, and we, the consumers who absorb it all indiscriminately—needs to reexamine the seriousness of the situation. We should admire the talented but only for their actual talents. We shouldn't fall victim to Hollywood's latest illusion—that they have something to tell us about the dangerous world we live in.

FIVE

FRANCE:

FROM GREAT TO INGRATE

Fifty-six thousand six hundred and eighty-one American troops lie buried in military cemeteries in France, many of them at the Normandy beaches, where the long straight symmetrical rows of crosses and Stars of David rise and fall with the terrain before the eye.

The United States and France need no treaty to bind them together. The ties are written in blood.

Yet France is suffering from a national case of amnesia, forgetting the obligations that come with the lives we have lost fighting for French freedom. During the war on terror and, in particular, the battle to free Iraq, they have been our adversary, blocking action any way they could.

OFF WITH THEIR HEADS!

French independence has often clashed with the moral obligations that history imposes. When President Charles de Gaulle demanded that the United States withdraw its troops stationed in France as part of the NATO force, American secretary of state Dean Rusk, in a rare foray into humor, asked darkly if de Gaulle meant to include the dead soldiers buried in French cemeteries.

Throughout the cold war, France flirted with becoming a "third force" between the United States and the Soviet Union, recognizing China well before Washington did and condemning America's war in Vietnam—which we inherited from the French.

But this latest assertion of Gallic independence came at too high an emotional price for Americans to take in stride. This time, the plea for French help came not from the civilized notes of diplomats but in a long plaintive wail from our hearts. Vulnerable in a world awash in terror, we asked our allies to rally around and help us out.

Britain, our best and most steadfast friend, was there for us. Battling alongside our troops in Afghanistan and joining our efforts in Iraq, Prime Minister Tony Blair paid no heed to the political consequences among the largely dovish British public and plunged full ahead in backing America. Our hearts beat more quickly when President Bush introduced him as he sat in the balcony during the president's speech to a joint session of Congress and to all of America pledging to destroy terrorism. Our friend was there in a time of need.

Others rallied, too. Spain, Portugal, Italy, Holland, and the other European Union countries rejected the appeal of France, Germany, and Belgium to oppose American efforts to disarm Saddam Hussein by force. When war with Iraq finally came, troops from old friends like Australia and new ones like Poland battled alongside British and American units.

But Germany was not there. Berlin broke its postwar tradition of backing U.S. initiatives, and Chancellor Gerhard Schroeder won reelection campaigning against war in Iraq. We bemoaned German forgetfulness. Where was their gratitude for the American airlift that saved Berlin in 1948, when the Soviets blockaded the city? What of the hundreds of thousands of American troops stationed in Germany to protect its liberty during the cold war—sixty-nine thousand of whom are still there?

But German intransigence was somewhat forgivable. Germany was not so much opposed to *this* war but to all war. The suffering of the past century has created a broad antiwar German constituency, an understandable and, in truth, comforting fact. After all, who can really object to pacifism in a former adversary who twice plunged the world into war?

But France was a different story. Where Germany's disagreement came despite their general reluctance to diverge from Washington's position, France's opposition to the war seemed to be an explicit result of the fact that the United States was the protagonist.

Russia also opposed the American insistence on forcing Saddam to disarm—but, again, that may not be as surprising: After all, Russia has been an adversary, not an ally, for nearly all of the past century (except for the years during World War II). China objected as well, but few had expected this Communist nation to come to our aid.

What we did expect was that France would stand by our side. They did not. Instead they stood behind us, plunging in the knife.

But appeasement, and the perfidy it breeds, are not new to the French soul and psyche. They have dwelled there—sometimes virulent, sometimes in remission—for almost a century.

Jacques Chirac, the French president, saw political advantage at home and abroad by defying the United States and appeasing terrorism. He has led his nation into an abyss from which it will not soon recover. Turning his back on his traditional friends and alienating their people, he has allied himself instead with tenuous replacements.

Chancellor Schroeder in Germany, his main confederate, has a popularity that has sunk to almost minus numbers. President Vladimir Putin in Russia wants and needs the United States a lot more than he does France. China will always act in its own narrow self-interest and has no friends, only temporary partners.

For the affections of these, Chirac has thrown away two hundred years of friendship and turned his back on one hundred years of moral indebtedness.

OFF WITH HIS HEAD!

Why Does France Act the Way It Does?

Some have pointed to financial motives in sizing up France's desire to placate Iraq. The reason is clear: Saddam Hussein had pledged to France the right to develop two of his four vast new oil fields—if the U.N. embargo should ever be lifted. But while the resulting cash windfall might seem large, by the time war with Saddam became inevitable, it must have dawned on the French government that America would never let Saddam stay in power—and thus they couldn't possibly count on the oil deal ever coming to pass. . . . Money, while an obvious explanation for the French position, doesn't fully explain why Paris has been so determined to oppose the war: France almost certainly had more to gain than lose financially by siding with Britain and the United States. After all, who was more likely to win, the coalition forces or Saddam Hussein?

The deeper answers lie in the bowels of French politics. The Muslim population in France, a leftist neo-Communist hatred of the United States, anti-Semitism in the national psyche, and a desire by even political moderates to distance themselves from American influence all play their roles. But the dominant factor is a generic desire to appease, rather than confront, that has dominated French thinking since her massive losses in World War I.

Ever since France relinquished its colonies in the Arab world—Algeria, Morocco, Tunisia, Syria, and Lebanon—it has become the destination of choice for Muslim immigrants seeking to better themselves economically. Between 4 and 5 million Islamic immigrants have made their homes in France—about 7 percent of the French population, proportionately speaking, more than half as large as the Hispanic population in the United States.

With so heavy a domestic Islamic presence, France has to be careful not to antagonize Arab nations, and French politicians must take the Islamic voting strength into account.

Beneath the veneer France presents to the outside world, though, lurks a darker and harsher reality of French politics.

Racism on the right and communism on the left are ever-present factors in the French electorate. In the most recent presidential election, 17 percent of France's voters backed Jean-Marie Le Pen, a racist candidate running on a nativist platform. Anti-Semitism is rife among his supporters. Another 17 percent backed the candidates of the Troskyite Workers Struggle, the French Communist Party, or the ultra-left-leaning French Green Party. Altogether, 34 percent backed extremists of the Left or the Right.

So, to begin with, one French voter in three is nuts.

Le Pen, who finished second in the French presidential election, once "dismissed the Holocaust as a 'detail' of history," according to the Reuters New Service. His vote was a chilling reminder of the anti-Semitism that lies just beneath the surface of French politics.

Indeed, it was not until more than half a century after the Holocaust that the government in Paris "officially admitted that France helped the Nazis persecute Jews during World War Two."

Until then, the version of France portrayed in the movie *Casablanca* had dominated our consciousness—the patriotic French rising defiantly to sing "La Marseillaise" as their German masters scowled. Yet a truer portrait of the French might include a reminder that their authorities helped the Nazis round up seventy-five thousand Jews for deportation to death camps in Germany.

The six hundred thousand Jews in France today—the largest Jewish population in Europe outside of Russia—have suffered increasingly severe acts of anti-Semitism and violence:

- A thirty-four-year-old rabbi was stabbed after a threat was made against his life by anti-Semitic groups.
- French schools recorded 455 anti-Semitic acts in the first trimester of the 2002–2003 academic year; French education minister Luc Ferry has called anti-Semitism a "true danger."
- A synagogue in Marseilles was burned to the ground, and three others were attacked, on March 30–31, 2002.
- A car bomb rammed a synagogue in Lyon.
- In Montpellier, a Jewish religious center was firebombed.
- A synagogue in Strasbourg was also bombed, as was a Jewish school in Créteil.
- A Jewish sports club in Toulouse was attacked with Molotov cocktails.
- In Bondy, fifteen men beat up members of a Jewish football team with sticks and metal bars.
- The bus that takes Jewish children to school in Aubervilliers has been attacked three times this year.

And all in 2002–2003!

In total, French police have reported more than four hundred anti-Semitic attacks between autumn 2000 and spring 2002.

And these dramatic episodes of violence only scratch the surface of the more casual, socially accepted anti-Semitism that runs through French culture. *Dreaming of Palestine,* a book written by a teenager glorifying suicide bombings in Israel, has climbed to the best-seller lists. The governing body of the Pierre and Marie Curie campus of the University of Paris asked the European Union to suspend ties with Israel, saying that such exchanges backed Israeli policy in the Mideast.

Israeli military action against the Palestinian terrorists has brought out the latent anti-Semitism in the French psyche. After Israel launched a reprisal raid on the West Bank on March 29, 2002, synagogues in France were torched and Jewish graves desecrated. Shimon Samuels, of the Nazi-hunting Simon Wiesenthal Center, blamed French premier Lionel Jospin for "inaction."

When the racist Le Pen finished second in the French presidential race and entered the runoff against Chirac, Europe was appalled, but Samuels felt the roots of Le Pen's triumph lay deep in the passive French response to the anti-Semitic outbursts. "Had they [the French government] taken measures," he said, "that could have been a valuable tool against what we saw tonight."

Israeli prime minister Ariel Sharon warned that French Jews were "facing a dangerous wave of anti-Semitism," and Israel's deputy foreign minister, Michael Melchior, said that France was "the West's worst country for anti-Semitism."

The close identification of American and Israeli foreign policy leads French anti-Semites to back the Saddam Husseins of the world against the George W. Bushes. Sympathy for the Israel-hating Arab world is broad and widely felt in France.

France also has its demons on the extreme left. Between one-fifth and one-quarter of the electorate generally votes for Communist candidates or for their derivative parties. Deeply anti-American and devoutly anticapitalist, most of these leftist voters find the official Communist Party too conservative and Stalinist; instead they have flocked to the Trotskyite Struggle Party.

Opposition to American world designs, and strong backing for the Palestinian cause, are deeply rooted in the French ultraleft, which had a viable chance of being elected in the 1950s. With the extreme left still controlling much of the French labor movement, this stridently anti-American element is a serious power to be reckoned with in French politics. This extreme left views Israel as imperialistic, threatening third-world Arab nations through its technological superiority, funded by a flow of American aid. Never mind that Israel is a democracy—to the French left wing, it is the enemy.

Part of the power of the French left comes from its credibility during World War II, when it was the only viable resistance movement in France. Though General Charles de Gaulle got all the credit for his Free French movement, he stayed safely tucked away in London while the largely Communist resistance battled it out in the streets with the Nazis. The taint of collaboration with Germany never quite left the French Right, and it empowered the left in the postwar era.

With about one-third of France voting for the extreme left or the racist right, it takes only a modicum of support from sane and normal voters to swing French politics in crazy directions.

But beyond right-wing anti-Semitism and leftist anti-Americanism, the tendency to appease dictators and agressors runs deep in the French psyche.

The Roots of French Appeasement

Appeasement—the tendency to give an aggressor what he wants, in the hope that he will go away and leave you alone—is ingrained in the French political soul and spirit. In the years before World War II, France tried to appease Hitler, giving in to one after another of his warlike moves.

France will often seek the easy way out. This reluctance to confront stems not so much from any cowardice in the French *esprit* but from a deep antimilitarism that has its roots in the carnage of World War I. As David Gelernter has noted in his *Weekly Standard* article "The Roots of European Appeasement," "Modern Europe's visceral loathing of war is a consequence of World War I." In four years, between 1914 and 1918, France lost one-quarter of its young men; many, many more were seriously and permanently wounded. At the single battle of Verdun, fought over a few miles of territory, hundreds of thousands of French soldiers died. In all, 1.45 million Frenchmen lost their lives in World War I—the cream of the youth of a nation of 45 million people. By comparison, American combat deaths in World War I amounted to 136,516 soldiers and in World War II to under 300,000. In the undeclared war in Vietnam, the loss of fifty-eight thousand Americans left a huge scar on our national psyche. It is difficult for Americans to appreciate the enormity of the French loss of life during WWI: A comparable level of U.S. casualties in a war today would see 10 million dead American soldiers.

In World War I, Europe was one big line of trenches, hundreds of miles long, parallel to one another in a sickening symmetry from the English Channel to the Ardennes Forest on the German border. In these wet, diseased, fungus-ridden hellholes, millions of French and British soldiers waited to hear the whistle that would send them charging toward an enemy trench, across several hundred yards of no-man's-land strewn with barbed wire.

Hypnotized by the logic of the offensive, Allied generals hurled their young against machine guns emplaced in the German trenches, whose deadly efficient fire cut them down. The French achieved no surprises in their attacks, since they insisted on preceding each foray with hours of artillery bombardment that usually did little good and only opened up craters in the no-man's-land, which became foxholes where terrified men could hide.

Without the element of surprise to worry about, German reinforcements were quickly shuttled up or down the defensive trench line to the point of attack by interior railroads. With the offense advancing on foot and the reinforcements arriving by rail, it was never a fair fight. Not until the invention of the armored tank, late in the war, did the offense begin to work again as a military tactic.

Not that the Germans were any better. As soon as the Allied offensive would end in a flood of blood, the Germans would attack, with the same deadly results, for the same obvious reasons.

No matter how often the offenses failed, they kept on coming, piling up more and more dead. When it eventually became clear that Britain and France would win the war, it was because their combined populations, and therefore their potential armies, were larger than that of Germany. When the United States added its population to the allied manpower reservoir, the Germans gave up.

Every French town, no matter how tiny, has a plaque honoring the dead of World War I. I recently visited the village of Souillac, in the Dordogne region of France. A hamlet of four thousand people, the war memorial immortalized the names of 158 young men who had died between 1914 and 1918.

The mindless slaughter of the war left the French with a strong aversion to combat and a deep lack of confidence in the military resolution of international differences. France's premature capitulation to the Germans in World War II (after only a six-week invasion) can be traced directly to this memory of annihilation in the first war.

Back then, the idea of "appeasement" had not yet developed the stigma it possesses today. In the years before World War II, both the French and the British proudly used the word to describe their policy of dealing with Adolf Hitler and the Third Reich.

The sordid story of appeasement began in the early 1930s, when France refused to react to German rearmament under the Nazi Party—a flagrant violation of the Treaty of Versailles that ended World War I. Barred by treaty from developing an air force, Hitler rearmed in secret, while France was content to take his official denials at face value.

Sound familiar?

Next, Germany publicly refused to honor the war reparations demanded of it in the Versailles agreement. No reaction from France. When Germany announced openly that it was no longer going to abide by the Versailles armament restrictions, France accepted the news and took no action. Hitler marched his army into the demilitarized Rhineland area, the heart of German industry, in defiance of the Treaty of Versailles . . . but once again the Allies did nothing. Then he threatened to occupy the Sudetenland, the area of independent Czechoslovakia where most of that nation's German population lived. French and British leaders hurried to a conference with the führer in Munich—and emerged having agreed to give Hitler the territory in return for his pledge of no further aggression. Neville Chamberlain, British prime minister, returned to England, umbrella in hand, promising "peace in our time."

But Hitler was undeterred by the flimsy pledge; he invaded the rest of Czechoslovakia soon afterward and then began to threaten Poland. Finally, drawing a line in the sand, Britain and France announced that an invasion of Poland would mean war with them as well. When Hitler attacked again, World War II finally began.

Appeasement is a philosophy, founded in fear and uncertainty, in which contradictions abound and live alongside one another. Then and now, French appeasement has been characterized by both a refusal to believe anything bad about an aggressor and a fear of what he might do if riled. Both in the 1930s and in the Iraq crisis, the French appeasers came to dislike their allies more than those who are threatening the peace.

In the 1930s, France consistently maintained that Hitler was acting merely to rectify the injustices of the Treaty of Versailles. Looking back with a measure of guilt at the harsh terms its leaders had imposed on Germany, France was quick to justify Hitler's refusal to pay reparations or abide by armament restrictions as merely an understandable rebellion against a draconian treaty. Gelernter describes how "the victorious allies soon came to feel that the peace they had dictated to [Germany] ... was vindictive and unjust—especially the huge reparation payments imposed on Germany as punishment for having started the war."

When Hitler moved to annex neighboring nations, French politicians cited the large number of ethnic Germans living there. These explanations of Hitler's aggression find their echo in the modern French insistence that Saddam Hussein was not concealing weapons of mass destruction and in Paris's sympathy for Saddam's assertions of sovereignty in the face of allied and U.N. intrusion.

Before World War II, many in France refused to believe that Hitler was a serious threat, while, at the same time, fearing what would happen if he were. France decided to hide behind a supposedly impregnable line of artillery and fortifications along the German-French border, famously called the Maginot Line. When Hitler did invade France, he simply walked around the line, leaving its turrets and guns untouched. Even today they sit there, pointed impotently eastward, symbolizing the folly of self-delusion.

France also derides the idea that Saddam is a threat, parroting his claim that he has destroyed his weapons of mass destruction. And yet, paradoxically, France also says she is worried that an allied invasion will prompt Saddam to use those same weapons. Like the folly of the Maginot Line, the French rely on appeasing Saddam to win his loyalty, hoping against hope that he and the other Islamic terrorists will therefore spare Paris in their global crusade.

As the clock ticked down to World War II, France discovered an enemy against which she could muster all of her emotional vitriol—not Hitler, but Great Britain. Henri Beraud, editor of the anti-Semitic French newspaper *Gringoire,* spoke for much of his nation when he wrote, "I hate England. I hate her by instinct and by tradition." To hear some French at the time speak, it was *Britain,* not Hitler, who was threatening their national essence. First the British were too timid. Then they were too militant. Then they were both. London was the bete noire of Paris.

As the threats posed by terrorism held the world in their thrall, Paris again found enemies: the United States and Israel. Rather than expend psychic energy opposing the terrorists, France devoted her attention to battling those who were fighting terror. This confusion of allies and adversaries is a hallmark of appeasement—an international variant of the famed Stockholm Syndrome, which leads hostages to see their captors as friends and their rescuers as enemies.

The Stockholm Syndrome refers to the results of a study of four hostages taken in a bank robbery in Stockholm, Sweden, on August 23, 1973. The Reverend Charles T. Brusca describes the findings: After the victims were held in a bank vault for five and a half days, "even though the captives themselves were not able to explain it, they displayed a strange association with their captors, identifying with them while fearing those who sought to end their captivity. In some cases they later testified on behalf of or raised money for the legal defense of their captors."

Initially, hostages come to identify with their captors as a defensive mechanism: The victim, in "an almost childlike way," tries to win the favor of the captor so he will not be harmed. He comes to fear that efforts to release him may expose the hostage to a renewed risk of violence, from the authorities or from the captor.

Then, Brusca continues, "the captive seeks to distance himself emotionally from the situation by denial that it is actually taking place. He fancies that 'it is all a dream,' or loses himself in excessive periods of sleep, or in delusions of being magically rescued." Hostages in such situations may even reverse the moral dynamic in their minds, becoming convinced that the criminals are in the right and the would-be rescuers are the problem.

Nothing could better describe the French reaction to terrorism:

- Trying to win the approval of the terrorist so as to minimize the chances of harm
- Raging against rescuers because of a fear of the by-products of their violence
- Engaging in busy work (like inspections) to keep their minds off the threat
- Denying that the situation is really taking place

Appeasement is also marked by internal dissent and factionalism. Divided, left from right, one ethnic group from another, a victim-nation cannot find within its soul the unity to confront an aggressor. Unable to launch a united defense, its various factions can only agree on appeasement as a response.

France in the 1930s was filled with such divisions. As the war loomed, the Socialist/Leftist Popular Front (including the Communists) took power, sending French business reeling. With over 1 million Jews in France and anti-Semitism rife, the French nation was hopelessly divided.

Much the same kind of division marks French policy today. With 4 million Muslims and six hundred thousand Jews, and large numbers of voters for both the racist right and the Communist left, France is a nation divided against its own interests. Today as in the 1930s, the only policy that seems able to unite the French people is a policy of inaction.

But appeasement didn't work in the 1930s, and it won't work today.

Stephen R. Rock, in his timely book *Appeasement in International Politics,* explains its defects as a strategic framework. (Intriguingly, he summarizes his argument with a French quote, "*L'appétit vient en mangeant*": "appetite grows with eating.") Rock says that the psychological impact of appeasement is the opposite of what is needed to deter aggression. "Appeasement gravely weakens the credibility of deterrent threats. Once it has received inducements, the adversary refuses to accept the possibility that the government of the conciliatory state will later stand firm. It thus advances to new and more far-reaching demands. When the government of the appeasing state responds to these demands by issuing a deterrent threat, it is not believed. Ultimately, deterrence fails, and the appeasing state must go to war if it wishes to defend its interests."

Or at least persuade the Americans to do it for them.

The French Attitude Toward America

Most countries ask only to be left in peace, to develop and govern themselves democratically. But some countries, conditioned by prior centuries of dominance, ask for more. Britain, Russia, Japan, Germany, and Spain have all learned harshly what happens when their empire is gone, and they are forced to retreat within their national borders. France, too, has had to shed its once vast colonial empire.

But the hegemony of France was never primarily political or military. The French ruled the world culturally. French wines, champagne, cooking, perfume, fashion, art, literature, political thought, philosophy, language, diplomacy, and customs have all at one point or another dominated the world. Indeed, the very phrase "Western civilization" has come to be seen as a synthesis of British political traditions and French culture and language.

(On the other hand, when asked what he thought of Western civilization, Mahatma Gandhi quipped, "It would be nice.")

As recently as the nineteenth century, France dominated global culture. French was the language of diplomacy, literature, science, and history. The czarist court at St. Petersburg, Russia, spoke not Russian but French!

But now the culture that dominates the world is, for better or worse, Anglo-American. English is the universal language. A Frenchman finds his culture overshadowed by American movies, television, music, fast food, magazines, and books. The global epicenter for serious cooking has arguably shifted to California, influenced by Asian spices, while Italian and California wines increasingly equal or exceed French in quality and sales.

An American might ask, "Who cares?" But to the French this shift is, indeed, climactic. Culture matters intensely to the French, in a way it never has, and likely never will, to the more practical Americans. The French feel overshadowed and overwhelmed, their very national essence compromised by the surge in American pop culture and the English language.

The generosity of the United States toward France only exacerbates the feeling of domination. British historian Arnold Toynbee once wrote that the United States "is a large, friendly dog in a very small room. Every time it wags its tail, it knocks over a chair."

French anti-Americanism has its roots in the political philosophy of General Charles de Gaulle. Taking power in Paris after the war and again in 1957, de Gaulle was, as Richard Bernstein writes in *Fragile Glory, A Portrait of France and the French,* "determined, after the humiliating defeat at the hands of the Germans in 1940, to make France felt in the world again. . . . What better way to make it felt than to snub its nose, occasionally, at the very great power that had restored its freedom and independence?"

Since the days of de Gaulle, French presidents have felt anti-Americanism to be "an imperative of genuine French independence." De Gaulle's successor, President Georges Pompidou, said France "was fated to be the 'emmerdeuse' of the world," a French word that Bernstein translates as "pain in the ass."

The famed nineteenth-century German chancellor Otto von Bismarck was quoted, by former Secretary of State Henry Kissinger, as noting that "the weak grow strong through effrontery and the strong grow weak because of inhibitions." Nothing could better describe the relationship of French and American foreign policy.

As Bernstein describes it, French anti-Americanism "has many sides, most of them having to do with an instinct to compensate for the natural limits of [their] country, the limits of population, national strength, the capacity to project power and influence beyond one's borders."

But throughout the post–World War II era, the cold war disciplined French intransigence, confining it to relatively innocuous forms. At its most offensive, France refused to allow American jets the right of passage over its air space in our 1986 retaliatory strike punishing Libya for its role in international terrorism. Such largely symbolic gestures infuriated Americans but did little to weaken us. The Soviet Union's shadow loomed over Europe, and despite its independent nuclear weaponry, France needed the United States as badly as she had in both world wars.

With the end of the cold war, however, France has graduated from being a mere "pain in the ass" to a serious obstacle to American foreign policy in battling terror. Now, all the strands have come together: Anti-Semitism, cultural resentment of the United States, political anti-Americanism, appeasement, and fear of terrorism have all conspired to drive the French into opposing and alienating their former American allies.

France: Saddam's Ally, Not Ours

France and Saddam Hussein have always been close. When it comes to the Middle East, Paris is no mere wayward ally of the United States, flirting with Iraq. The opposite is true: France has been a steadfast ally of Iraq, eager to support Saddam in staving off the United States.

France's flirtation with the Arab world, it must be said, is recent. In the years immediately after World War II, France sided with Israel against the militant Arabs. Still a colonial power in Africa and the Middle East, the French opposed and feared Arab nationalism, worried that it would cost them their colonies, primarily in North Africa. Locked in a bitter, bloody war to keep Algeria French, Paris wanted nothing to do with Islamic independence movements.

France was also anxious to play down the impression that it had collaborated with its Nazi occupiers during the war, so in order to show its antipathy to anti-Semitism, Paris strongly supported Israel in its disputes with the Arabs. Indeed, in 1956, French troops joined British and Israeli units in attacking Egypt to reoccupy the Suez Canal, which Cairo's dictator Gamal Abdel-Nasser had seized from the Anglo-French company that had run it.

But the situation changed radically in 1957, when Charles de Gaulle came to power and quickly granted Algeria its independence. Eager to mend his relations with the Arab world, de Gaulle ended his nation's longtime support for Israel and set Paris on a decidedly pro-Arab course. Ten years later, the pro-Arab tilt had become so evident that de Gaulle ordered a halt to French military aid to Israel—a week before the Arab nations attacked the Jewish state once again.

Saddam Hussein's first visit to a Western country came in 1972, when he agreed to sell oil to Paris. Two years later, as French prime minister, Jacques Chirac visited Baghdad to cement the relationship.

The love blossomed. France became a huge arms supplier to Iraq—second only to the Soviet Union—and, in 1976, helped Baghdad build a nuclear reactor at Osirak, which gave Saddam a huge leg up in his efforts to acquire nuclear weapons. So dangerous was the reactor that Israel bombed it in 1981, destroying, for the moment, Iraq's nuclear program.

But Saddam's ambitious efforts to expand his power had only just begun. In 1980, with France supplying many of the arms, he invaded Iran, starting an eight-year war that drained Iraqi lives and wealth. Paris supplied Saddam with Mirage F1 fighters, Exocet antiship missiles, and equipment to improve the accuracy and range of his Scud missiles as the war dragged on. One year before the Gulf War between Saddam and the world started, French defense minister Jean-Pierre

Chevenement visited Baghdad and announced that he wanted to improve the Iraqi-French relationship to what he said would be a "higher level."

But the war with Iran was expensive, so Saddam went deeply into debt with his Arab neighbors. When Kuwait, the small, oil-rich kingdom on Saddam's southeastern border, pressed him for repayment, he responded by invading on August 2, 1990.

Outraged at the violation of international law, all the Western nations—even France—rallied to Kuwait's side and launched Operation Desert Storm to roll back the Iraqi invasion.

French president François Mitterrand immediately dispatched French troops and ships to join the coalition against Saddam. But he also played his own diplomatic game. CNN reported that "while mobilizing militarily, however, Mitterrand also demanded that every diplomatic means possible be employed to resolve the crisis, a conflicting message that sowed doubts about French resolve and infuriated her American and British allies."

"Until the last minute we wanted to avoid war," former French foreign minister Roland Dumas said. He personally offered to open negotiations with Iraq in exchange for a promise to withdraw from Kuwait. "Everything was ready, but at the last moment," Dumas relates, "as we did not receive the promise from Saddam Hussein, I said I am not going to Baghdad." Angered at the invasion of Iraq, Defense Minister Jean-Pierre Chevenement resigned in protest. It was only after the French Embassy in occupied Kuwait was raided, and four French citizens kidnapped, that Mitterrand agreed to be tougher with Saddam.

Throughout the Gulf War, France remained ambivalent about the invasion of Iraq. Reporting in the *Washington Post* during the war, Jim Hoagland explained the French view at the time: "France, with 4.5 million residents of Arab origin, appears to feel that a quick, decisive military strike, followed by new Western efforts to convene an international peace conference to resolve Israeli-Palestinian differences, is the best course. Fear that an American military operation could be too successful from the standpoint of France's commercial interests generates some public opposition to the gulf operation, officials in Paris acknowledge."

Hoagland quoted a senior French official as saying: "There are those who fear that America is simply entrenching itself in the oil countries of the gulf for the next half century and will leave no room for us to operate there. That of course would be a war aim we could not go along with." The French official urged that after a successful invasion, allied troops should stop "at the [Kuwait] border and hope that the defeat in Kuwait and an international arms embargo would be enough to contain Iraq."

After the end of the Gulf War, France continued to play a pro-Iraq game, slowing and hampering U.S. and British efforts to hobble Saddam's regime and deny it weapons of mass destruction. The motive was commercial. On March 31, 1995, *USA Today* reported that Iraq owed France $5 billion—which helps explain why Paris opened a diplomatic-interests section in Baghdad later that year, a step the other allies refused to take.

The *London Mail* said that Western intelligence services had learned that French companies had signed multimillion-dollar contracts to help rearm Iraq, among other things, in exchange for oil.

The French cooperation with Iraq was paying off. The *Jerusalem Post* quoted Iraq's oil minister, Amer Rasheed, as saying: " 'Friendly countries who have supported us, like France and Russia, will certainly be given priority' " when the lucrative contracts for the reconstruction of Iraq are awarded after the oil embargo is lifted.

After the end of the Gulf War, the United States and Britain refused to lift sanctions on Saddam, even though he had been forced to disgorge Kuwait. In view of Iraq's record both before and after the Gulf War, the Allies realized that we needed to keep pressure on the Iraqi dictator not to develop weapons of mass destruction.

After all, he not only had such weapons, he had already used them. In the 1980s, Iraq bombed Iranian troops with poison gas—the first battlefield use of the weapon since the Germans deployed it against British troops in World War I. Saddam also attacked the Kurdish town of Halabja with mustard gas, killing hundreds. As CNN reported, "According to Physicians for Human Rights, trace elements of the nerve agent sarin were discovered after an [Iraqi] assault on the village of Birjinni."

After the Gulf War, with U.N. inspectors crawling all over the country, Saddam had to be more careful. For years the inspectors found very little, but then a breakthrough came—via a tip from Saddam's own family! "In August of 1995, Saddam Hussein's two sons-in-law—

together with his daughters—defected. One of them, Hussein Kamel, had been in charge of Iraq's secret weapons concealment operations. He started to reveal the inner workings of the Iraqi armaments program as soon as he arrived in Jordan."

He told the U.N. inspectors: "We were ordered to hide everything from the beginning. And indeed a lot of information was hidden and many files were destroyed in the nuclear chemical and biological programs. These were not individual acts of concealment, but they were the result of direct orders from the Iraqi head of state."

The brothers-in-law will have to share a posthumous Nobel Prize for Idiocy: Lured back to Iraq, Saddam had them shot.

How did France react to the discovery of Saddam's vast new weapons system? It began to press for *lifting* the embargo against Iraq, a step that U.N. Resolution 687, which set up the inspection system, made clear was only to happen after Saddam had destroyed his weapons of mass destruction.

Step by step, the French battled to win Saddam's favor and increase his strength and freedom to operate. Ignoring the evidence of his venality, Paris embarked on a course of pro-Saddam activity. A French delegation even visited Baghdad to push for reopening of commercial relations. A Paris foreign ministry official said he wanted "a policy of getting out of the crisis [with Iraq] and the sooner the better."

Despite Iraq's violations of the U.N. resolutions and the evidence of its full-fledged weapons program, France opened an "interests section" at the Romanian Embassy in Baghdad, headed by a senior diplomat—in effect extending diplomatic recognition to the regime.

In 1996, France signaled its increasingly pro-Saddam tilt, by opposing U.S. cruise missile attacks against Iraqi positions in the northern Kurdish part of the nation and by pulling out of the aerial surveillance operation in place since the Gulf War.

Meanwhile, Saddam moved brilliantly to develop support on the U.N. Security Council by auctioning off Iraq's oil to three nations with vetoes in the international forum—France, Russia, and China. In quick succession, starting in 1997, Iraq signed deals with Russian companies to develop the West Qurna oil field, with Chinese firms to drill the Al-Ahdab field, and with France's Total and Elf Aquitaine to develop two new fields in its rich southern territory.

But as the *Financial Times* noted: "These deals, however, can only go into effect and help develop the Iraqi industry when the sanctions are lifted. Not surprisingly, Russia, China and France have tried to push for an end to sanctions, leading to severe splits among the five permanent members of the U.N. security council."

Now that Saddam had lined up his allies in the United Nations, he was prepared to move. First he evicted U.N. inspectors in late 1997, claiming they were American spies. The French, playing their end of the tag team, used the crisis to call for a lifting of the embargo, saying that Saddam had to see light at the end of the tunnel. With France leading the way, the United States offered what Madeleine Albright called "a small carrot" in order to get Saddam to let the inspectors back in—she agreed to let Saddam sell more oil. Albright's "carrot" got the inspectors back in.

But by 1999, Saddam had thrown them out again, this time for keeps, and the United States and Britain had responded impotently with four days of bombings and missile strikes, which did nothing at all to force the Iraqi dictator off course. Getting the inspectors out was all that mattered to him; he would have taken forty days of bombings if he had to.

How did France respond? Unbelievably, they proposed giving Saddam the reward the United Nations was reserving for his cooperation— the lifting of sanctions. "French diplomats describe their proposals as a 'contribution' to get round the deadlock caused by last month's bombing and Baghdad's defiance of Unscom," the *Financial Times* reported.

Here's how the reasoning went: U.N. inspectors could no longer operate in the poisoned atmosphere created by the U.S. and British air strikes, so wouldn't it be better to lift the sanctions and rely on unspecified "monitoring" of Iraqi activities to deter arms violations? Freed of the embargo, they claimed, Saddam would open his arms to the inspectors he had just evicted.

The appeasers all gathered around the French proposal. Kofi Annan, U.N. secretary-general, praised France's suggestion as "a first step" toward solving the impasse. Iraq said that it "sees a need for a balanced dialogue to find practical solutions to the situation."

Of course, this plan flew in the face of the U.N. resolutions adopted after Desert Storm, which specified that the "oil embargo on Iraq can be lifted only once inspectors report that Iraq is free of its chemical, biological and nuclear weapons, and the long-range missiles used to deliver them." Thereafter "an ongoing monitoring system would be established once Iraq is disarmed to check that it isn't rebuilding its weapons programs."

But France didn't want to wait for Saddam to comply and disarm. It proposed that the United Nations "immediately shift UNSCOM's [the arms inspectors] work from active disarmament work to monitoring Iraq's known weapons sites."

Nobody explained how the "monitoring" was going to deter Saddam from developing further weapons of mass destruction now that he was being given a free pass on the ones he already had. Nor did France offer a reason to suspect that Saddam would cooperate at all once he had succeeded in lifting the only sanction that remained in place—the oil embargo.

France continued its appeasement of Iraq right up to 9/11. Shortly after Bush took office, Iraq fired on American and British planes patrolling the "no fly zone" in the north of Iraq, established to protect the Kurds from further harassment and attack after Saddam Hussein had gassed and killed them by the tens of thousands.

But starting in 1998, shortly after its oil deal with Iraq, France stopped participating in the patrols, "insisting that it failed to see any point in the exercise." After the Iraqi attack on Allied aircraft, France, Russia, and China called for an end to the patrols. A French official explained their point of view: "What the three countries have been saying is that it's clear that the patrols do not invite Iraq to co-operate" with the continuing process of controlling Iraq's aggressive behavior.

"Invite" Iraq to cooperate?

The choice of words alone leaves no doubt: If the French think a murderous dictator deserves the courtesy of an invitation to anything but unconditional surrender, it's clear they've ceded whatever moral authority they may once have claimed.

Whatever the roots of French ingratitude, it is maddening. Their national refusal to understand the threat to the Western world in general—and their longtime American protectors in particular—after the assault of 9/11, and their unwillingness to come to our aid, have seriously undermined French prestige in the United States.

A recent Fox News/Opinion Dynamics poll showed almost half of Americans supporting a boycott of French products in light of Paris's refusal to work with the United States to tame Saddam Hussein.

Will the relationship recover? The more important question is: Should it?

It's time for France to grow up and make choices. The almost infantile policy of acting out against the parental United States is a luxury the French may conclude they cannot afford if they are forced to confront the consequences of their own actions. The American people should not be anxious to forgive French ingratitude or to explain it away. Paris must understand that in opposing the United States in the war on terror, it alienates not only the American government but the people as well.

Diplomats move onto new diplomacy, politicians to new politics. But the American people are not so facile in their affections and resentments. Animosity against the Communist government in China, a result of Beijing's conduct during the Korean War, was so intense that American presidents dared not even establish diplomatic relations with the world's largest nation for twenty years thereafter. Resentment against the powers of Europe after World War I was so profound that it kept the United States out of the League of Nations, destroying that body's potential effectiveness.

France has so deeply alienated the American people that she has put herself behind the eight ball in global affairs. Americans have peered into the French soul and are revolted at what they see. They will not soon return to gawk at the Eiffel Tower or luxuriate in expensive meals at Parisian restaurants. Just as the French government should not expect an easy return to their old roost of American protection if they should encounter some future threat to their own security, the French people should not expect the steady flow of American tourist cash to be what it once was. They have broken our faith, and it won't be easily restored.

But a deeper question is how to treat the United Nations. For forty years the United States did not take the world body seriously, because its power was blunted by the Soviet veto on the Security Council. There was no point in asking the United Nations to intervene when the Soviet Union or one of its surrogates attacked a free nation, for Russia's veto could always be counted on to kill the motion.

(The sole exception was in 1949, when North Korea, backed by China, invaded South Korea: As it happened, the Soviet Union was boycotting Security Council meetings at the time, and with the Taiwan Nationalist Chinese regime still representing China, the United Nations was able to achieve the unanimity necessary for action.)

Now it is the French veto that has scotched the potential for U.N. action in the short term. But without the United Nations and without the Security Council, who cares if the French agree or not when the United States and Britain consider international action? With negligible military and economic power, the two allies can safely disregard Parisian sensitivities and act as they and their real allies wish.

So the lesson of French intransigence over Iraq is to ignore France and the United Nations as long as France has its veto. Just as the United States would never consider asking the General Assembly for approval (because of the influence of undemocratic third-world and Arab countries), it should not feel obliged to seek Security Council consensus. Negotiate with Russia, with China, with the rest of Europe—but leave France alone. Let them stew.

PART II

OUR OTHER ASSAILANTS

Not all those who destroy our institutions fly airplanes into buildings or amass stockpiles of poison gas. Not everyone who terrorizes us is a terrorist.

But there are other villains on our shores—evildoers who, in their own ways, have done much to harm our people and damage our institutions.

Those who **attack our economy** forced us into a recession, as surely as Osama bin Laden. He crashed four airplanes—three into crowded buildings—and sent our travel, entertainment, airline, and insurance industries reeling. But there were others, in our government and private sector alike, who'd been hatching their own schemes for years, swindling investors and sending Wall Street crashing.

Flight 93 might have been aborted before it could destroy the Capitol building in Washington, but the incumbent congressmen who **attack our democracy** got there long before with their own plan to hijack our democratic elections. By partisanship, gerrymandering, and manipulating the system, they stripped us of one of our most basic rights—the ability to choose freely the men and women who represent us in Congress. By creating a kind of permanent insurance policy for incumbents of either party, they savaged our democracy in ways Osama bin Laden could only dream about.

The tobacco industry **attacks our children.** Each year they hook more than a million kids on their product and kill more people than Saddam Hussein's and Kim Jong Il's arsenals together might manage. Four hundred thousand Americans die yearly at the hands of their weapon of mass destruction—the cigarette. And the Tobacco Terrorists have their accomplices: state governors, who have taken money that could have gone to fund antismoking efforts, and used it instead to balance their precarious state budgets.

And meanwhile another group of scoundrels has been preying upon those among us who most deserve dignity and reverence. Around this country, nursing home owners **attack our elderly**. In their hellholes of abuse and neglect, they beat, torture, rape, taunt, demean, and humiliate those of our parents unlucky enough to fall into their grasp. With a brutality that Saddam Hussein could only admire, they are the terrorists of the old.

Those who attack us need to be brought to justice, their abuses brought to an end.

OFF WITH THEIR HEADS!

SIX

THE ATTACK ON OUR ECONOMY:

HOW TWO SENATORS—CHRISTOPHER DODD AND PHIL GRAMM—PASSED LAWS THAT HELPED ENRON DEFRAUD ITS INVESTORS WITH IMPUNITY

When those planes rammed into the World Trade Center, slicing our hearts open, Osama bin Laden was striking at our freedoms, our power, and our capitalist system. It's no accident that he chose the towers of Wall Street as his principal victims.

He succeeded. Not only did he bring down the office buildings, he also brought the American and global economy to a standstill. Since then we have suffered—not from a Bush recession but from the bin Laden recession.

But Osama had help.

In the boardrooms of America, those who have benefited the most from our free enterprise system had already hatched a whole host of plots and schemes designed to defraud investors and undermine the confidence that kept the economy growing. But these corporate directors and CEOs had confederates in their plans: accountants and lawyers who showed them how to do it and get away with it.

Furthermore, they couldn't have pulled it off without the help of politicians. Two senators in particular—Christopher J. Dodd, Democrat of Connecticut, and Phil Gramm, Republican of Texas—made billions of dollars in larceny possible. In the 1990s, when we weren't looking, they pushed through two laws that, in effect, immunized Wall Street from lawsuits by investors whom it swindled. These laws protected Enron and Arthur Andersen so they could cook the books in peace. These senators also helped to stop the Securities and Exchange Commission (SEC) from curbing some of the worst abuses on Wall Street.

Peel back the layers of the Wall Street onion, and what do you find? On the top layer are the corporate executives who committed the frauds. Next are the accountants who taught them how to do it. And at the rotten core are these politicians who passed laws to protect them from the consequences of their actions.

And these laws are still on the books!

Try as they might, the investors whose life savings are gone will be lucky to get pennies on the dollar back. Why? Because that's how Wall Street and Capitol Hill planned it.

OFF WITH THEIR HEADS!

By the time the dust of the Enron scandal had settled, tens of thousands of investors had lost billions, as the company's stock plunged from $90 to $1 in a few days. *Time* reported that more than half of the Enron employees' 401(k) assets, "or about $1.2 billion, was invested in company stock, which is now nearly worthless. Billions more were lost by other investors, from individuals to large institutions that bought Enron shares for the pension plans of unions and corporations."

If the poor suckers who bought Enron stock, or the energy company employees who had no choice but to purchase it, were stuck when the company tanked, the top executives made sure they came out fine. The *New York Times* noted that "as Enron stock climbed and Wall Street was still promoting it, a group of 29 Enron executives and directors began to sell their shares. These insiders received $1.1 billion by selling 17.3 million shares from 1999 through mid-2001."

Enron chairman Kenneth L. Lay "sold Enron stock 350 times, trading almost daily, receiving $101.3 million. In all, Mr. Lay sold 1.8 million Enron shares between early 1999 and July 2001, five months before Enron filed for bankruptcy."

Ken Lay got out in time, of course. But plenty of others didn't. William S. Lerach, a prominent securities plaintiff's attorney who is suing corrupt Wall Street firms, has called attention to the story of Roy Rinard, a fifty-four-year-old utility lineman employed by an Enron subsidiary. Roy was one of the unlucky ones: His 401(k) account, invested entirely with Enron, shrank from $472,000 to less than $4,000 after Enron declared bankruptcy. He was helpless to stop the loss. Why? Among other reasons, Rinard and other Enron employees were prevented by company rules from selling their retirement plan stock. Only top management had that privilege.

Beyond Enron, the crisis in the energy company set off a wave of reverberations, with corporate disasters hitting Global Crossing, World-Com, AOL, and a host of other companies that swamped investors. The shock waves are still being felt today on Wall Street, as investor confidence has sagged to lows not seen since the stock market crash of 1929.

What caused the crisis? How could Enron have gotten away with phony statements of profit and loss, false reports of earnings, and deceptive projections of its future in the closely regulated environment of publicly traded companies policed by the Securities and Exchange Commission?

Easy: Arthur Andersen, the major accounting firm, showed them how. It conducted what amounted to a private tutorial for Enron executives on how to lie and cheat. Meanwhile, Arthur Andersen's name, reputation, and imprimatur on the company documents guaranteed that the data they fudged would be accepted as accurate and fair.

It wasn't the first time that the Andersen firm had been caught lying about a client's earnings. The *Chicago Tribune* describes how the accounting firm paid out $110 million in 2001 to settle shareholder lawsuits in connection with the Florida Sunbeam Corporation. The lawsuit stemmed from accounting gimmicks that "pumped up" Sunbeam's earnings in 1997 by $70 million.

But the real question is how could Arthur Andersen help Enron misrepresent its data and hope to get away with it?

The Politicians Sell Out to Arthur Andersen

As so often happens, the answer goes back to politics—specifically, to a deal cut between the Democrats, led by Connecticut's Chris Dodd, former chairman of the Democratic National Committee, and the Republicans, led by Phil Gramm of Texas, in the 1990s. A deal with all the hallmarks of political double talk, it was fueled by massive campaign contributions from the accounting industry. It is a tale of a powerful industry's deliberate manipulation of the legislative process to pass laws that *hurt* the consumer rather than help him—that protect those who defraud the investor rather than punish them.

The story began on April 19, 1994, when the U.S. Supreme Court, in a particularly pernicious 5–4 decision, ruled that investors could no longer sue accountants who had vouched for phony claims of profits made by corrupt corporate executives. The familiar right-wing coalition of Justices William H. Rehnquist, Anthony M. Kennedy, Antonin Scalia, Clarence Thomas, and Sandra Day O'Connor said that the statutes regulating securities transactions did not permit those who had been defrauded to go after accountants or others whose actions were "aiding and abetting" the fraud.

Now, investors could sue the company that issued the statements (which was usually broke)—but not the accountants who had approved them.

The dissenters, led by Justice John Paul Stevens (and including Harry Blackmun, David H. Souter, and Ruth Bader Ginsburg), pointed out, "In *hundreds* of judicial and administrative proceedings in every circuit in the federal system, the courts and the SEC have concluded that aiders and abettors are subject to liability" under federal law. They bemoaned the majority's reversal of this practice, saying that the ability to sue aiders and abettors "deters secondary actors . . . from contributing to fraudulent activities and ensures that defrauded plaintiffs are made whole."

The Supreme Court ruling in 1994 set the stage for a brutal legislative battle in 1995. Faced with such a wholesale reversal of long-treasured investor protections, Congress felt bound to act to restore some of the rights the Court had stripped away. But soon the vultures honed in on the proposed remedial legislation, to make sure that the worst abuses—and abusers—would still enjoy the protections the Court decision gave them.

But the political landscape changed dramatically in the midterm elections of 1994. Clinton and the Democrats, who had controlled both the House and the Senate in the 1993–1994 session, now lost both chambers to Republican majorities. The GOP legislators were emboldened by their radical conservative agenda, enshrined in a "Contract with America" issued by future House Speaker Newt Gingrich to rally his troops for the decisive election of 1994.

The contract called for "common sense legal reform" to prohibit aider and abettor liability. It also wanted to limit the damages of those who were liable. "Under current law," it read, "a defendant can be held responsible for the entire award [stemming from a lawsuit] even if he is not completely responsible for all the harm done." The Republicans pledged to change things by assigning to each actor liability for only the portion of the damage he caused.

This theory *sounds* good—but the truth is that, in most securities frauds, the scam could never have been pulled off unless all the actors were on board. A crooked company needs a crooked accountant, and a willing lawyer, to make the fraud stick. If either one is honest and blows the whistle, the fraud doesn't happen.

So how much of the liability in a fraud case should by shared by a dishonest accountant who lets a dishonest corporate executive issue a false statement of profits, earnings, losses, and expectations for the future? If the executive refused to issue the numbers, they wouldn't go out; if the accountant refused to ratify them, they would have no credibility. So each is a necessary actor: You can't have a fraud unless they both play ball So they both should be fully liable. (Especially in cases where a corporation is long since bankrupt, unable to pay back the investors, while the accountant might otherwise walk away unscathed).

But the Republicans who took over the Congress in 1994 didn't see it that way. Led by Phil Gramm, they were determined to weaken investor protections, perhaps in the misguided belief that it would encourage enterprise and entrepreneurship in the national economy.

Meanwhile, though, another, more sinister actor—Chris Dodd of Connecticut—was planning to use the GOP impetus to further his own agenda.

Even before the 1994 Supreme Court decision, Connecticut had been rocked by a huge scandal that was a precursor to the Enron affair. The *New York Times* reported that in the early 1990s, six thousand Connecticut investors had been lured to invest in a firm called Colonial Realty by inflated reports of earnings. The investors lost tens of millions, and the scandal reached so far into the highest ranks of Connecticut's officialdom that, when the investors sued, the judicial bench had trouble finding a judge who hadn't lost money to try the case.

The optimistic predictions of Colonial Realty were endorsed by Arthur Andersen and by the law firm of Tarlow, Levy, Harding & Droney. Ultimately, Andersen was forced to pay some $90 million to settle the Colonial case. And Droney's firm had to come up with $10 million.

The "Droney" at the end of the firm name was John Droney, formerly Connecticut State Democratic chairman. Dodd nominated Droney's brother, Christopher, as U.S. attorney for Connecticut and later to the federal bench as U.S. district court judge.

When Dodd learned of the Supreme Court decision banning aiding and abetting lawsuits, he apparently thought of his friend John Droney, who was facing just such a suit for his handling of Colonial Realty.

Dodd then sponsored a bill to make sure that accountants, lawyers, and other professionals couldn't be sued for aiding and abetting the fraud. Amazingly, Dodd even sought to make the provision *retroactive* in what looked like a blatant attempt to shield those implicated in the Colonial Realty case.

The *Hartford Courant* noted that the "original draft" of Dodd's bill "put cases currently pending under his proposed law—such as the Colonial case." Eventually the retroactive provision of Dodd's bill was dropped, removing the protection to Colonial Realty. But he was still able to insulate his accountant and lawyer pals from the consequences of any future frauds.

Once the consumer groups learned what Dodd was up to, they protested vigorously, denouncing his proposed bill. Ralph Nader called Dodd's legislation "The Financial Swindler's Protection Act of 1995."

Dodd's efforts to protect his lawyer and accountant friends went much further. An article in the *Legal Intelligencer* reported that Dodd's bill "would set a minimum threshold of losses below which investors could not sue accountants and limit lawsuits to 'primary violators,' meaning that accountants could be sued only if they were directly implicated in wrongdoing."

Dodd was also eager to ensure that anyone who lost a lawsuit against his accountant and lawyer friends might face having to pay for their legal fees. Michael Calabrese, executive director of Public Citizen's Congress Watch, said that Dodd had "created a bill that's out of control and now has tremendous protections for the financial services industry."

Dodd's bill also limited the liability of accountants, lawyers, and other professionals to a portion of the losses caused by their fraud.

Perhaps the most scandalous feature of the Dodd bill was that it created a "safe harbor provision," which allows public companies to be shielded from litigation when their projections and predictions of future earnings and profitability turn out to be bogus. All they have to do is to put in what one financial adviser called "adequate cautionary language"—a disclaimer—and all is cured. "Lie all you want," the legislation seemed to provide, "just put in some boilerplate language and you'll be okay."

Dodd's bill also handcuffed lawyers trying to help investors to get their money back. The Consumer Federation of America pointed out that the bill "requires that a victim's complaint, filed at the beginning of the case, 'state with particularity all facts giving rise to a strong inference that the defendant acted with the required state of mind.'" Columbia law professor John Coffee calls this provision "a Catch-22: You can't get discovery unless you have strong evidence of fraud, and you can't get strong evidence of fraud without discovery."

Before the Dodd bill, investors could sue companies for civil violations of the Racketeer Influenced and Corrupt Organizations Act (RICO). Civil RICO has teeth. It allows for an award of triple damages and attorneys' fees. And what better way to describe the shenanigans that went on between Enron and Arthur Andersen than that it was a "corrupt organization?" Because civil RICO was effective, Dodd made sure that it was removed from the diminishing quiver of weapons with which an investor could protect himself. Under the bill, investors could no longer sue under the RICO statue to recover their losses.

Summing up the provisions of this terrible bill, William Lerach wrote that the "changes were a bonanza for public companies and their insiders, investment bankers, and financial accounting firms, i.e., the normal defendants in securities cases. Higher pleading standards, automatic discovery stays, a safe harbor that arguably permits corporate executives to lie about future results . . . damage limitations, elimination of joint and several liability for reckless conduct, and, for good measure, a mandatory sanction review procedure that . . . threatens plaintiff's counsel with up to 100% liability for defendants' fees."

The way forward for the Dodd bill was greased by massive campaign contributions to candidates for Congress—including Chris Dodd himself, who got $54,843 from Arthur Anderson alone, more than any other Democratic senator, and $37,750 from computer companies that prudently supported the legislation, which would protect them in case their projections went awry.

No wonder the *New York Times* called Dodd "perhaps the accounting industry's closest friend in Congress."

Overall, during the 1995–1996 campaign cycle when the Dodd bill was pending, the accounting industry and the big accounting firms gave $7,782,990 to congressional candidates.

But still, the bill didn't have an easy time of it. As consumer groups lined up against it, President Clinton came under enormous pressure to veto it. Within the administration, a fierce debate raged on whether to sign or kill the legislation.

The Phony Clinton Veto

As the president's pollster, I advised a veto, noting that public opinion strongly disagreed with the legislation. In a survey conducted in November 1995, voters overwhelmingly rejected the provisions of the bill.

Then I ran into Bruce Lindsey, the president's oldest friend and closest personal adviser. The venue for the encounter was an odd one: the men's room on the second floor of the West Wing. Lindsey asked me about the securities bill, and I said, "I advised him to veto it. The bill is terrible, and it'll make a great issue for us against the Republicans."

"A lot of Democrats favor it, too," Lindsey noted.

"Sure, but when has that stopped us?" I asked.

"Well." His tone turned serious. "We're getting a lot of pressure from our friends in California to sign it."

"You mean the Silicon Valley types?" I asked.

Lindsey nodded. The technology hub in northern California was a key source of support for the president and a big contributor to his campaign.

"The issue will do us more good than the money," I parried.

Lindsey shrugged, as if to say, "We'll see."

This conversation with Bruce Lindsey stands out in my mind because it was the only time, in my two years of work with Clinton in the White House, that I ever heard anyone mention a policy issue in terms of its effect on possible campaign contributions. Despite the pressure to raise money to fund our ambitious schedule of television ads, I never heard a single suggestion that we might change or alter any policy to get more money into our campaign—until the men's room conversation with Lindsey.

When I spoke to the president about the bill in early December of 1995, he explained his dilemma to me: "Not only is the Silicon Valley on me about the bill, but so is Dodd. He wants me to sign it," he said.

"You can't be pushed around by those guys," I responded. "The issue is too good for us. It will allow us to run against the Republicans as the folks who want to rip off old ladies and other investors."

"But what about Dodd?" the president persisted. As chairman of the Democratic National Committee, the Connecticut senator was a key member of the Clinton team and responsible for much of the fund-raising. To go against him on a matter of this importance could result in serious bad blood.

And Clinton didn't need bad blood with Dodd, certainly not then. In November and December 1995, the securities bill was an afterthought. Center stage was fully occupied by Clinton's resistance to the budget cuts the Republican Congress was pushing. Led by Speaker Newt Gingrich, the GOP had closed down the federal government after the president vetoed their package of harsh budget cuts.

Holding up Clinton's side of the argument was a $10 million program of television ads emphasizing why the president needed to stand firm "for America's values" and block cuts in "Medicare, Medicaid, education, and the environment." Without the media advertising, Clinton would never have been able to get his message out. Dodd—and the financial interests for which he was speaking—controlled a lot of the money we needed.

Clinton proposed a solution. "Let's do it like we did on the highway bill in Arkansas," he suggested.

He was referring to his political maneuvering, as Arkansas governor, when a bill was introduced by the highway construction lobby in 1983 to raise taxes on heavy trucks to fund highway construction and repair. Clinton was torn between the highway contractors, who were key financial supporters of any incumbent governor, and his own worry about raising taxes.

Clinton had had good reason to worry about road taxes. As a freshman governor in 1980, he had been defeated for reelection largely because he raised car license fees to fund road construction. He worried that if he signed a bill for the truck tax hike, he could be in trouble all over again.

Clinton solved the problem by trying to please both sides. First he satisfied the highway lobby by endorsing the bill. Then he doubled back and told the truckers he opposed it. The state highway director was less than pleased and called Clinton a "double-crosser." While Clinton ended up signing a watered-down bill, he had skirted a tough issue that could have hurt him politically.

"What if I veto the bill and it's overridden? Would the override hurt me politically?" Clinton asked. He'd yet to have a veto overridden by Congress.

"No," I conceded, "as long as you're forthright in opposing the bill and veto it, an override won't hurt you. The public doesn't care if you get overridden. They just want to see you fighting the good fight against the Republicans."

"Even if Democrats join in the override?" he prodded. What he meant was: Would people see through my veto if the Democrats vote to override me—would they realize the veto was just window dressing?

"No," I answered, "even if Democratic senators vote for the bill, that's their political problem. It won't interfere with your standing against it."

The die was cast. On December 20, 1995, Clinton vetoed the bill, saying, "I am not willing to sign legislation that will have the effect of closing the courthouse door on investors who have legitimate claims. Those who are the victims of fraud should have recourse in our courts. Unfortunately . . . this bill could well prevent that."

But even as Clinton was vetoing the bill, Dodd understood that he would incur no presidential wrath if he overrode the veto. So the Connecticut senator worked overtime to repass the bill, lining up the two-thirds majority he would need to make it law. Dodd, a loyal party man, would never have dared to override a Clinton veto if he hadn't been fully confident that the president wouldn't mind.

Reading the mixed signals from the White House and feeling pressure from their campaign contributors, the Democrats fell in line and voted to override their president's veto. Twenty Democrats joined the

Republicans in the Senate override, and eighty-nine Democratic congressmen voted to override in the House, joining an almost solid GOP vote for the bill.

Even smart consumer advocates seemed fooled by the Clinton two-step. They attacked Dodd, but they let Clinton alone. Charles Lewis, of the Center for Public Integrity, said, "Chris Dodd—here he is, chairman of the Democratic Party, but he's also the leading advocate in the U.S. Senate on behalf of the accounting industry, and . . . he helps overturn the veto of his own president, who installed him as Democratic chairman. Dodd might as well have been on the accounting industry's payroll. He couldn't have helped them any more than he did as a U.S. senator."

Lewis didn't get it. In effect, Dodd *was* on their payroll—through campaign contributions.

For his part, Clinton never let on that the whole charade had been prearranged and choreographed. He got credit for standing up for the consumer by vetoing the securities bill, while one of his chief fundraisers, Senator Dodd, could continue to rake in money for Clinton from Wall Street, the accounting industry, and the Silicon Valley as a payoff for passing it anyway.

Indeed, after Enron collapsed and hapless investors found they couldn't go after Arthur Andersen, Clinton sanctimoniously blamed the Republicans, saying he had vetoed the bill, which, he said, "cut off investors from being able to sue if they were getting the shaft." He said that he was "sure some of the people in Congress that stopped a lot of the reforms I tried to put through are probably rethinking that now."

That's chutzpah.

Andersen and Enron Get to Work Defrauding Investors

Once the securities bill had passed, Arthur Andersen could get to work helping Enron defraud investors without having to worry about lawsuits.

Here's how they did it:

"At the heart of Enron's demise," *Time* reports, "was the creation of partnerships with shell companies, many with names like Chewco and JEDI, inspired by *Star Wars* characters. These shell companies, run by Enron executives who profited richly from them, allowed Enron to keep hundreds of millions of dollars in debt off its books. But once stock analysts and financial journalists heard about these arrangements, investors began to lose confidence in the company's finances. The results: a run on the stock, lowered credit ratings and insolvency."

Why did Enron and Arthur Andersen decide not to own up to the debts these shell companies were racking up? They were protecting Chief Financial Officer Andrew S. Fastow, whom the *Times* described as "the driving force" behind the phony accounting procedures. "Evidence introduced at the criminal trial of Arthur Andersen indicates . . . that [an] improper accounting decision—which set in motion Enron's destruction—served mainly to benefit the financial interests of a single corporate insider. . . . While the decision brought few if any benefits to Enron itself, these accountants said, it did help to protect the financial health of an outside partnership managed by the company's chief financial officer then, Andrew S. Fastow."

Despite this evidence of malfeasance, investors cannot sue Arthur Andersen for their losses with any hope of significant recovery—because of the protections Chris Dodd got passed in the Securities Litigation Reform Act of 1995.

Blocking Separation of Auditing and Consulting

But the Securities Act changes weren't the only service Dodd and his colleagues rendered to the accounting industry. As the 1990s unfolded, one of the most honest men in Washington—SEC commissioner Arthur Levitt Jr.—began to worry about the integrity of the audits of the major accounting firms. Concerned that these firms had a conflict of interest in auditing companies (like Enron) with which they also did consulting business, Levitt sought to bar accounting firms from consulting for companies they audit.

The principle seemed fair enough. An auditor must be free to speak out against false numbers and to demand corrections in the published financial statements of public companies—but if these same auditors are getting huge consulting fees from their clients, they might be reluctant indeed to kill the golden goose.

As it happens, that is just what went on between Enron and Arthur

Andersen. As Enron's auditor, Andersen was expected to be objective and impartial. But the firm was heavily dependent on consulting fees from Enron. (In 2001, for example, Andersen was paid $27 million by Enron for consulting services and $25 million for its audits.) Hiring such a firm for this kind of double duty is a bit like hiring your IRS agent as your personal accountant: He'd inevitably be torn between his desire to collect taxes from you, and his wish to continue to get your fees for his accounting services.

Levitt—whom the *Washington Times* describes as "one of the most aggressive SEC chairman on behalf of investors ever"—wanted accounting firms to stop consulting for companies they audit. He "was convinced audits were being compromised because the firms were protecting their consulting business."

Worried, according to the Associated Press, that "accounting firms are jeopardizing their independence by becoming more financially dependent on the lucrative consulting work they do for companies they audit," the SEC chairman campaigned to separate the two in the closing months of the Clinton administration. The *Washington Post* describes how he worked "feverishly . . . crisscrossing the country from Dallas to New York for meetings while juggling a blizzard of calls and visits to members of Congress." His proposal "sparked a firestorm of protests" from accountants, led by the American Institute of Certified Public Accountants.

USA Today reported that thirty-eight congressmen and fourteen senators, most of them members of the oversight committees with jurisdiction over the SEC, called Levitt to urge him to back off. Chief among them were Representative Billy Tauzin (R-La.), Senator Chuck Schumer (D-N.Y.), and Senator Phil Gramm (R-Tex.).

Tauzin, chairman of the House Energy and Commerce Committee, had received $143,424 in campaign contributions from the accounting industry in the preceding five years. He wrote to Levitt that he saw "no evidence" of a problem justifying the SEC action.

Schumer had taken $329,600 from the accounting industry over the last five years. He wrote to the SEC opposing the rule change, a letter SEC officials said "was almost certainly composed with the assistance of the accounting lobby." After the Enron scandal broke, Schumer donated $68,800 he had gotten from Enron and Arthur Andersen to a fund for former Enron employees. He says that he defended the accounting industry not because of the campaign contributions but to protect thousands of jobs in New York City.

But nobody was as compromised in his actions, or as influential, as Texas Republican senator Phil Gramm, then chairman of the Senate Banking Committee. The *Washington Times* reported that Gramm wrote the SEC questioning whether there was any evidence that accounting firms were "cooking the books" or "looking the other way." He also said that the proposed SEC rule change would "force dramatic changes in the structure and business practices of accounting firms" and require corporations "to pay increased costs for some types of accounting services."

Gramm, who may have quit the Senate in 2002 to avoid having to defend his Enron record on the campaign trail, is a special case in compromising relationships. Gramm's wife, Wendy, sat on the Enron Board of Directors and on its Audit Committee, for which she was paid $22,000 annually plus $1,250 for each meeting she attended. (*Frontline* reported how she was "named to the company's board, just five weeks after stepping down [as Chairman of the Commodities Futures Trading Commission] which around the same time exempted Enron . . . from federal regulation on some of their commodities trading . . . a big financial boon to Enron.")

Wendy and Phil got out in time. She sold all her 10,256 shares of Enron stock for $276,912 on November 3, 1998—for $27 per share, considerably above the $1 it would plunge to two years later.

Chris Dodd joined the fray of those pressuring Levitt. The Associated Press reported that he "helped broker a deal between the Securities and Exchange Commission and the Big Five accounting firms, which ended the SEC's push to restrict auditors from selling consulting services to their clients." The deal was, in reality, a surrender by Arthur Levitt.

Now accounting firms were freed to audit the same clients they consulted for—the conflict of interest that led directly to the Enron/Arthur Anderson scandal. For accountants to turn in their corporate clients for cooking the books would entail biting the hand that fed them.

But the special interests still had more dirty work for their hired hands on Capitol Hill to do.

1998: The Rape of Investors Continues

The 1995 securities law barred the doors of the federal courthouse to those who sued accounting firms to get back their life savings. Before long, investors began to respond by suing in state courts.

So, in 1998, Congress passed a law barring the state court route, too.

Attorney James E. Day, an associate at the law firm Kirkpatrick & Lockhart, noted that the 1995 act "by placing procedural and substantive obstacles to prosecuting securities class action litigation in federal court, led to an increased number of such suits being filed in state courts under state law." The special interests couldn't stand that, so, "spurred by evidence . . . of this 'noticeable shift in class action litigation from federal to state courts' Congress passed SLUSA [the Securities Litigation Uniform Standards Act] to promote the federal courts as the uniform forum and federal law, namely [the 1995] Reform Act, as the uniform standard governing most securities class action litigation."

In other words, having stacked the deck against plaintiffs in the securities litigation, Congress proceeded to ensure that state courts could offer no relief.

The Conference Committee reporting out the bill in Congress, in effect, said the same thing. "The solution to [the problem of the increase in state court securities class actions] is to make Federal court the exclusive venue for most securities fraud class action litigation involving nationally traded securities."

After the 1998 act passed, Day explained, any investor who complained about "an untrue statement or omission of a material fact in connection with the purchase or sale of a covered security" couldn't go into state court, but had to litigate in federal court—where he could not hold accountants liable for the frauds they permitted.

The new bill was passed on July 23, 1998, explicitly at the behest of the Silicon Valley companies. The *Tech Law Journal* was frank in relating how the bill was "designed to decrease the number of harassment suits brought in state courts that threaten the ability of companies—particularly high-tech Silicon Valley companies—to raise capital and disseminate information."

Congressman Rick White (R-Wash.) said that "our thriving high technology companies need protection from frivolous lawsuits that prey on their volatile stock prices. This bill will help those companies focus their energies on the marketplace instead of the courtroom, and keep them providing the innovative products and services we have come to expect."

The bill limited pretrial discovery, forced plaintiffs to contend with the "safe harbor" defense for phony projections passed in the 1995 act, and permitted the high-tech companies to survive the collapse they faced in 1999–2002—all without being exposed to lawsuits in state courts.

Clinton signed the bill, signaling how phony his veto of 1995 had been: Now here he was, signing a bill to stop investors from circumventing the same rules he had previously vetoed.

Part of the reason Clinton didn't veto the bill but felt he had to sign it, of course, was his growing political weakness. In the interim, the Monica Lewinsky case and the impeachment that ensued had weakened his always-limited ability to defy the special interests and the call of his party's senators to give them what they wanted.

Representative Bart Stupak (D-Mich.) correctly observed, "If we pass this bill, Congress will place all investors into a largely untested, untried new federal system that will make it very difficult for investors to prove fraud." How right he was.

So now—after the 1994 Supreme Court decision in the Denver case, the congressional passage of the 1995 Securities Litigation "Reform" law, the stymieing of Arthur Levitt's efforts to ban consulting and auditing by the same accounting firm, and the 1998 Securities law—the investor was delivered, bound and gagged, over to the fraud mavens at Enron, Arthur Andersen, Global Crossing, and a host of other companies.

The stage was set for the massive failures and frauds of the early 2000s.

The Phony Reforms of 2002

Once the bombs had exploded, Enron had failed, WorldCom had gone up in smoke, Arthur Andersen had closed its doors, and confidence in Wall Street had sunk to the Elton John level—too low for zero—Congress and the Bush administration acted. Just in time for the midterm elections of 2002, Congress passed and Bush signed the Corporate Reform Act of 2002.

The bill included needed changes in rules for accountants, including the ban on auditors consulting for companies they audited, for which Arthur Levitt had fought. It included a number of important reforms, which certainly made sense:

- Accountants would be regulated by a new board under the SEC.
- Auditors would have to rotate every few years.
- Companies could not make loans to their directors or executive officers.
- CEOs would have to sign financial reports saying that they fairly present the financial condition of their companies, with criminal penalties if they lie.
- Directors or executive officers of a company would have to observe the same blackout periods on sale of their stock that employees have to observe in the pension plans.
- All off-balance sheet transactions would have to be disclosed.

But nothing in the legislation rolled back the efforts of the 1990s to hamstring investors seeking to get their money back. In the aftermath of the Enron collapse, Senate Democrats tepidly explored whether to reverse the horrendous bills passed in the previous decade, but nothing came of it. Congress wasn't willing to take away the special protections it had given those who defrauded investors—not when they also gave so generously to their campaigns.

The spin artists at the White House had deflected the corporate scandals, turning them into a law-and-order, cops-and-robbers spectacle, featuring corporate executives being led away in handcuffs.

As *Newsweek* put it: "Around the jail it's called a 'perp walk,' . . . cops parading a newly arrested 'perpetrator' in handcuffs or other heavy-metal wear past the waiting cameras. It's a mean-streets tactic viewed with disdain by the lordly federal prosecutors of the U.S. Attorney's Office in the Southern District of New York, especially in white-collar cases, where the perps wear suits and have connections."

But Bush needed a perp walk. With the scandal about corporate abuses threatening to tarnish his image and that of his party, a high-profile arrest would do his ratings good. So the administration focused on the case of John Rigas and his family's Adelphia Communications company.

When Rigas, accused of looting his company, was led away in handcuffs, under the gaze of cameras assembled for the purpose by the White House, Bush had his symbolic show of toughness. "Wait'll you see what's next," joked White House adviser Karl Rove. "Orange jumpsuits!"

Newsweek explained: "The Rigas arrests were only one part of an all-out White House effort to, as they say in the spin-doctoring business, 'get out ahead of the story.'"

In the legislative debate, congressmen and senators vied with one another to impose ever-tougher theoretical penalties on corporate executives who misrepresented their company's finances. The final law imposed a maximum ten-year sentence for a "knowing" violation and a twenty-year term for a "willful" one.

But all of this, of course, was nothing more than show and window dressing. Nothing was done to enable those who had been defrauded to see a dime of their money or to restore the only real threat that could discipline the business community—the overhanging risk of litigation by disgruntled stockholders.

As long as the enemies were the bureaucrats or the regulators, corporate executives understood that campaign contributions to their bosses could nullify their efforts. Helpless when their elected public officials jerked their leash, these enforcers could be kept under control. It was the investor, unrestrained by political ambition and empowered by access to lawyers eager to make a big fee, of whom they needed to be afraid. So the crippling legislation of 1995 and 1998 remained on the books, unchanged.

What We Need to Reform the Process

The Consumer Federation of America has issued a sensible plan to correct the abuses that caused the corporate scandals of 2001–2002. It's so sensible that it will never pass—unless the American people focus on it and get behind the legislation.

Among the measures it calls for:

Get Rid of the Safe Harbor

The safe harbor protections are like the papal indulgences that caused the Reformation. "Sin all you want—just put in a disclaimer," they say. We need to stop letting accountants and corporate executives hide behind fine-print disclaimer language when they make phony predictions about their companies. Go back to the old standard, before it was watered down by the 1995 law; predictions must be made in "good faith" with a "reasonable basis"—no caveats, no excuses.

Hold Aiders and Abettors Fully Responsible

Anyone who enables fraud should be responsible for its consequences. Under the "Reform" laws of 1995 and 1998, accountants can shut their eyes to fraud, even show executives how to *commit* fraud, and then say, "Who, me?" when the fraud is uncovered.

Re-impose Joint and Several Liability on Accountants

And, once the fraud is discovered, make the accountants, auditors, and other professionals fully liable for the fraud they cause—not just for a small part of it.

Make It Possible for Investors to Win in Court

Undo the rules of the 1995 law, which require that a victim of fraud know all the details before he or she can begin the suit. Give them the power to investigate, through discovery, while they are suing.

Permit Investors to Sue Under Civil RICO

If these Wall Street conspiracies between corrupt corporate executives and equally corrupt accounting firms aren't "corrupt organizations" within the meaning of the RICO act, what are they? We need to restore the ability of investors to sue under civil RICO when they've been fleeced by these experts.

Let Investors Sue in State Courts

Republicans love states' rights . . . until they get inconvenient (as they did in counting the votes in the 2000 election). Repeal the Securities Litigation Uniform Standards Act of 1998, to let investors sue in state courts where the deck may not be so stacked against them.

.   .   .

Wall Street hasn't been the same since the corporate fraud scandal. Investors are voting with their feet to stay away from the markets, until they can persuade themselves, and their families, that the system works. Like gamblers who have been fleeced by loaded roulette wheels, they're staying away from the tables until they decide the game isn't fixed.

Believers in the free-market system, investors are prepared to take a licking from time to time—as long as their losses are based on truthful accounts of a company's finances and on reasonable projections about its future. When a firm like Arthur Andersen permits a firm like Enron to lie, who can count on anything a corporate executive or his auditor says? Until and unless the Congress and the White House realize that it's this fundamental sense of unfairness that's holding investors away from the markets, they won't see the return of the bull market anytime soon.

All the measures the government has passed to "reform" Wall Street have left out one thing: redress for those who have been screwed. Where can they go to get their money back? To class-action lawsuits? That'll net them pennies on the dollar. To arbitration before Wall Street–appointed judges? Securities lawyer Robert Weiss puts it best: That route is "rigged for the Wall Street houses." Jury trials? Almost every investor had to sign away the right to sue when he signed up with a brokerage company.

We must act quickly to grant special relief to those who have lost their savings to make them whole.

Without this guarantee, investors are on strike. And they should stay out until real reform is adopted.

SEVEN

THE ATTACK ON OUR DEMOCRACY:

HOW INCUMBENT CONGRESSMEN AND THEIR POLITICAL BOSSES TOOK AWAY OUR POWER TO CHOOSE OUR HOUSE OF REPRESENTATIVES

There is no antitrust law for politicians. When the leaders of both parties get together in a conspiracy, there are no statutes that permit zealous Justice Department lawyers to close in and prosecute.

But while Osama bin Laden was trying to destroy American democracy from the outside in 2001, the Republican and Democratic Parties did a pretty good job that same year of doing it from the inside. Their bosses and incumbent congressmen got together and redrew all the district lines for seats in the House of Representatives, with one goal in mind: to guarantee the reelection of all incumbents.

Democratic congressmen got districts filled with registered Democrats. Republicans got all the GOP voters. It was like a double wedding: Both parties got hooked up with districts that would remain faithful so long as they both shall live. The parties would still fight over vacant seats, but the politicians got what they were after: They took the *representative* out of "House of Representatives."

This incumbent protection program took the ballot out of our hands and guaranteed lifetime seats to 90 percent of all congressmen, regardless of what we thought of the job they were doing.

One by-product of this gerrymandering was that Democrats virtually conceded control of the House of Representatives to the Republican Party, in return for the safety for all their incumbent congressmen. Republicans happily obliged, drawing the lines in such a way as to guarantee that they would control the House for the next decade.

To Republicans and Democrats alike, then, it looked like a win-win situation. The only losers were us—the voters. The framers of our Constitution had designed the House of Representatives, elected every two years, to be the branch of government most susceptible to public opinion. Now, the incumbents of both parties have torn up the framers' plans.

Columnist Richard E. Cohen described the result best: "Even the Communist Party, during its heyday in the former Soviet Union, often faced more uncertainty" than the incumbent American congressmen who ran for reelection in 2002.

OFF WITH THEIR HEADS!

Here's how the deals worked.

In **California,** the Democrats used their control over both houses of the state legislature and the governorship to draw district lines to reelect all their incumbent congressmen (except Gary Condit, whom they tossed to the wolves as a sacrificial offering to the Media Gods). The Republicans didn't mind, because they got to protect their people as well.

So while $68 million was being spent in the tightly fought governor's race in the nation's largest state, not one of its fifty-three congressmen (besides Condit) lost the election. Not one.

With Hispanic voters accounting for 80 percent of the state's population growth in the past decade, the politicians had to work hard to keep Latino politicians out of Congress. But they did it well. Governor Gray Davis made sure that white, Anglo, Democratic congressmen Bob Filner of San Diego and Howard Berman of the San Fernando Valley kept their seats, by cutting Hispanics out of their districts. As the *New York Times* reported, "Governor Gray Davis and the Democratic Legislature, who controlled redistricting, redrew the lines of those districts to include fewer Latinos and, thus, protect Mr. Filner and Mr. Berman from the threat of a Latino primary challenge."

Now these two hypocrites—Filner and Berman—can continue to posture and strut in Congress about fighting for Hispanic rights—after they kept their seats by so dividing the Hispanic vote that it didn't get the representation it deserved in Washington.

Thomas A. Saenz, vice president for litigation of the Mexican American Legal Defense and Education Fund in Los Angeles, was not amused: "What they have done is split the Latino community to protect incumbents, and we think there are serious problems with that."

The political mapmakers did their job well. The five incumbent California congressmen who had tough races in 2000 (elected with less than 55 percent of the vote) all increased their margin substantially in 2002. Three got more than 70 percent of the vote in their new, friendlier districts. And two others got more than 59 percent of the vote. The California legislature took good care of its incumbents.

As the Center for Voting and Democracy put it in their report *Monopoly Politics:* "California is the poster child for efforts to shield incumbents. . . . How did they do it? By methodically dividing voters so that potentially vulnerable incumbents received more partisan votes from neighboring districts whose members either didn't need them to win or who didn't want them."

Dan Schnur, a Republican strategist, noted that "if the average Californian doesn't like his congressman, the only option is to call the moving vans." (Not for the congressman, mind you, but for the voter!)

Typical of the shenanigans in the nation's largest state was the sweet kiss given Representative Ellen O. Tauscher, a Democrat who was elected in 1996 by only four thousand votes. The *New York Times* reported that "California's redistricting plan shifted a substantial number of new Democratic rural constituents into her district" and Republicans "basically conceded the race." Tauscher was unopposed by any Republican.

Political heaven. Tauscher describes her reaction to the good news that she had no opposition: "My staff and I were standing in my kitchen and afraid to open a bottle of champagne because we were afraid someone made a mistake." In the old days of American democracy, of course, winning candidates had enough respect for the voters to hold their champagne until election night.

Another California congressional district—the twenty-third—was so creatively drawn by the state's political bosses that it was described by *Congressional Quarterly Weekly* as "skinny as a snake." The magazine noted that at one point, the district thinned "to span the distance between the ocean and the high tide line."

The goal of this abstract expressionist art was to reelect Democratic representative Lois Capps, who had survived three competitive races in a "slightly Republican-leaning" district. Her district had moved "well inland from the coast to take in conservative, rural areas." But now she has been blessed by the party bosses with a district that "takes in coastal cities and Hispanic enclaves in its 220-mile stretch through San Luis Obispo, Santa Barbara, and Ventura counties north of Los Angeles."

How did the Democrats let the Republicans keep control of the House of Representatives? Just look at California. They drew districts for their incumbents that were loaded up with every Democratic voter they could find, guaranteeing their members lifetime incumbency— while making it almost impossible to pick up new seats in other districts, now denuded of Democrats.

Congressional Quarterly Weekly describes what happened: "Rather than press for greater gains in [California], Democrats shored up all seven of their [vulnerable] House incumbents. . . . In the process, they guaranteed nearly all of California's incumbents safe seats until the redistricting that will follow the 2010 census."

The *New York Times* reported: "To the dismay of some national party strategists, state officials took a conservative approach," protecting all the Democratic incumbents. "While the state redistricting plan gave Democrats one new seat and eliminated the seat of a Republican . . . who has announced he will retire, the plan largely shored up existing House Districts."

In **Florida**, Gerrymander Jeb, the president's brother and the state's governor, worked with his legislature to draw districts so partisan that Republicans now outnumber Democrats in the delegation by 18–7— despite all recent memory of the famously even split between the two parties in the 2000 presidential election.

Florida gained two members of Congress in the reallocation of seats following the 2000 census. Not only did the Republican politicians ensure that both went for the president's party, but they also shored up their vulnerable incumbents. As *Congressional Quarterly* noted, "Florida may have trouble devising a user-friendly ballot, but state lawmakers had no problem crafting a redistricting plan that virtually guarantees the reelections of most House incumbents."

Again, as in California, the more Republican the legislature made certain districts, the more Democratic they made others, in their shameless effort to shore up incumbents.

As a result of the gerrymandering, seven Florida congressmen faced no major party opponents in either their primary or general elections. The beneficiaries of this bipartisan deal included five Republicans and two Democrats. Generous to the last, the legislature also made sure that Democratic state senator Kendrick Meek, who sought to succeed his mother (a five-term congresswoman), had so partisan a district that no Republican ran against him.

But Republicans weren't above a little poaching to pad their lead in America's third most populous state. Their chief target was Democratic representative Karen L. Thurman, who had won five House elections, most by wide margins, until the Republicans gerrymandered her district to exclude the heavily Democratic city of Gainesville and to include more of Florida's Republican west coast north of Tampa. The reapportionment helped Republican state senator Ginny Brown-Waite unseat Thurman in one of the few turnovers in the 2002 elections.

But in gerrymandering Florida, Jeb got an assist from brother George. The new congressional lines in southern states had to get federal approval under the Voting Rights Act (a requirement that wasn't imposed on northern states, which allegedly lacked the South's history of racial discrimination at the voting booth). When the Florida legislature submitted its plan to the Justice Department for "preclearance," the *New York Times* reported how the Washington lawyers hastened to approve it quickly to "undermine a main element of a Democratic court challenge" to the new lines that favored the GOP.

The Justice Department wasn't so obliging when **Mississippi** submitted its reapportionment plan, which reconfigured the state's districts to compensate for the loss of one congressional seat. That plan, endorsed by the state's Democratic governor and its black community, would have endangered one of the state's Republican congressmen, Charles W. Pickering Jr. (son of Bush's controversial nominee to serve on the federal Court of Appeals). So the Bush Justice Department took its time in reviewing the plan. When Justice failed to act, the federal court did and invalidated the proposed district lines.

Thus, as a result of "action by a federal court and inaction by the Justice Department" (in the words of the *New York Times*), the new Mississippi map forced Democratic incumbent congressman Ronnie Shows to run against Pickering in a district the Republican could not lose. The black voters who supported Shows when he represented the Mississippi delta were diluted by Republicans, and Shows lost by almost thirty points. This blatant use of the powers granted to the Justice Department in the Civil Rights Act to defeat a candidate supported by most of the African Americans in his district is an insult to all Americans.

In **Georgia,** Democratic leaders leaned over so far trying to fix the election results that it blew up in their faces. The chance to capture the two new seats the state had won in the 2000 census proved an irresistible temptation.

The centerpiece of the Georgia Democratic plans was the fatherly wish of party boss and Senate Majority Leader Charles Walker Sr. to deliver a congressional seat to his thirty-four-year-old son, Charles Walker Jr., by cramming as many Democrats into the district as possible. When he had finished his mapping, he had assured his boy's victory—or so he thought—by giving him a 60 percent Democratic district.

But Junior blew it. With only one year of college under his belt and four arrests (for leaving the scene of an accident, shoplifting, driving with a suspended license, and interfering with a police officer), even the solidly Democratic district rejected his candidacy.

The Democrats in Georgia were absolutely without shame in the way they gerrymandered their state's congressional districts. Republican state representative Lynn Westmoreland said the Democratic leaders had only one goal—to draw "six districts with [at least] 54 percent Democratic performance" in previous elections. "In many cases . . . strips of land were incorporated in districts" in order to pick up small sections of black population to bolster voting strength in Democratic districts. Westmoreland pointed to a bridge of land that he said is "700 feet wide and two miles long to connect two former districts."

Democratic leaders redrew the district lines in secret and revealed their finished maps only on the very day they passed them. But the Georgia Democratic reapportionment was so partisan that voters rebelled. Ralph Reed, the former Christian Coalition leader who became head of the Georgia Republican Party, cited voter resentment against the reapportionment as a key factor in the GOP victories. "There's no question about the fact [that] the gerrymandered redistricting that the Democrats passed in 2001 backfired. They created a lot of discontent at the grass roots."

As Westmoreland said, "The Democrats have controlled Georgia for 130 years, yet it took them only sixty days to carve Georgia up into what looks like a war zone."

Democrats paid, big time, for their high-handedness. They lost the U.S. Senate seat held by Democrat Max Cleland and the governorship. They also lost not only the Walker seat but another district as well, where Republican state senator Phil Gingrey defeated Democrat Roger Kahn—"despite," as *The Hill* pointed out, "the legislature having redrawn the district in an attempt to keep it in Democratic hands."

How do you get a state that voted for Gore in 2000, as well as Clinton in 1992 and again in 1996, to elect eleven out of nineteen Republican congressmen? Call your friends the mapmakers.

Pennsylvania Republicans were so ruthless in the gerrymandering that the U.S. District Court ruled that their reapportionment plan was unconstitutional. But the Republicans not only gerrymandered the state, they deliberately took their good sweet time doing so.

The result? By the time the maps reached the court, the statewide May 21 primary was approaching, and the Democrats were caught between a rock and a hard place. If they litigated against the Republican plan, it would mean a postponement of the primary date. This, they agreed, might fatally handicap their candidate for governor, who was locked in a tough primary race. As a result they threw in the towel and agreed to the Republican plan—a plan that had already been ruled unconstitutional.

Republican governor Mark Schweiker giddily signed the reapportionment bill. The lines were final; the result preordained. As one GOP operative gloated to columnist Richard Cohen: "Democrats . . . are in a hell of a box. . . . They are whistling past the graveyard if they assume they will get fair treatment."

Virginia's Democrats had a serious problem: how to reelect a crook. The Honorable James P. Moran, congressman from Virginia, is a lot less than honorable. His scrapes with the law began in 1984, when he served on the City Council in Alexandria, Virginia. Accused of helping a developer buddy of his win a bid for public land, he pled guilty to a conflict-of-interest misdemeanor to avoid a felony charge for vote peddling that was hanging over his head. He got a year's probation and had to quit the council.

But after a time in exile, Moran got elected to the House of Representatives in 1990 and began to distinguish himself in Washington. In 1995, he threw a punch at California Republican Randy Cunningham and yelled "I'll break your nose" at Indiana Republican Dan Burton during a public hearing. Off the playing field, he put a choke hold on an eight-year-old boy he accused of trying to hijack his campaign car in 2000. On June 23, 1999, his wife summoned the police to their home during a domestic disturbance. No charges were filed, but she sued for divorce the next day.

Moran's most recent scrape came in January 2001, when he accepted a $25,000 personal loan from lobbyist Terry Lierman, who represented the drug company Schering-Plough. Columnist Michelle Malkin noted in the *Washington Times* that "after getting that unsecured loan at a lower-than-market interest rate, Mr. Moran cosponsored a bill that would extend the patent on Schering-Plough's allergy medicine Claritin—and prevent generic drug manufacturers from offering inexpensive alternatives."

Last year, the *New York Times* reported that Moran "has been entangled in conflict-of-interest accusations involving his personal finances and his support for legislation to make it tougher for other people in financial trouble to declare bankruptcy."

In 2003, Moran got in even more trouble for saying that Jewish influence was the key factor in our decision to attack Iraq.

A political consultant's worst nightmare.

But none of this was any problem for the Virginia Democrats, who drew a can't-lose district in what the *New York Times* reported was a "deal with state Republicans" that "kept his district firmly Democratic." Moran won easily; on election night, he probably slept like a baby.

So egregious was the redistricting in **Michigan** that the federal court went out of its way to call the plan unfair, noting in its opinion that "despite the increasing majority of Democratic voters in Michigan, Republicans are likely to win ten of Michigan's fifteen congressional seats under the challenged plan."

But unfair wasn't unconstitutional, in the view of the three federal judges. The Constitution requires only that the districts be equal in population. Gerrymandering generally *isn't* unconstitutional—that's how Congress gets away with it. It's just unfair. In the end, the GOP did come out ahead—but only by a 9–6 margin. A small—*very* small—victory for democracy.

The bosses in **Connecticut**, Democratic and Republican, agreed on one thing: They didn't like Congressman Jim Maloney. Too conservative for the Democrats (he voted against tax increases) and too liberal for the Republicans (he's prochoice, pro–gun control)—he was an embarrassment to the orthodox leaders on both sides of the aisle. Maloney was just too independent for his own good.

When census takers told Connecticut that it would have to lose a seat in Congress, both party's leaders cast hungry eyes on Jim Maloney's chair. They paid him back for his independence by putting him in a district with twenty-year incumbent congresswoman Nancy Johnson to fight it out. And they made sure it was Johnson who would win the uneven contest, by filling the new district with more of Johnson's voters than Maloney's. Did the bosses punish Maloney? Not that anyone could prove. But everybody knew exactly what was happening, and why.

The House of Representatives Becomes as Democratic as the House of Lords

One would have thought the election for the House of Representatives in 2002 would have been a brawl. With the chamber almost evenly divided after the cliff-hanger election of 2000, even a switch in a few seats would have delivered the body to the hands of the Democrats.

Indeed, as the 2002 elections loomed, the parties were about as close to even as one can get. Democrats controlled about one-third of the state governments, the Republicans one-third, and the rest were split. The *New York Times* reported: "Most polls show[ed] near-even party identification, and a tie on party preferences for Congress."

Generally, the elections of 2002 were, indeed, close and narrowly fought. Just as America split evenly between the two contenders for president in 2000, so the contests for senator and governor throughout the nation were tight and tense. Twenty of the thirty-six governorships up for grabs changed parties. Four of the twenty-eight senators who ran for reelection lost their seats, and fourteen of the winning Senate candidates squeaked by with less than 55 percent of the vote.

But the House of Representatives changed hardly at all. The outcome was all preordained. Ninety-six percent of the incumbents who ran for reelection won, and 90 percent of the winning House candidates coasted to victory with more than 55 percent of the vote. While governor and Senate candidates sweated out election night 2002, most House candidates could have gone to bed early and slept peacefully through an uneventful night.

This wasn't how the founding fathers planned it. They envisioned the House of Representatives as the chamber of Congress most fully representative of the opinions of the people, reflecting popular whims, ideas, moods, and even prejudices. They counted on the more sedate Senate, with its six-year staggered terms, to mitigate the often impulsive electoral decisions that would shape the lower chamber.

In the *Federalist Papers* James Madison wrote, "The House of Representatives, with the people on their side, will at all times be able to bring back the Constitution to its primitive form and principles. Against the force of the immediate representatives of the people, nothing will be able to maintain even the constitutional authority of the Senate, but such a display of enlightened policy and attachment to the public good as will divide that branch of the legislature the affections and support of the entire body of the people themselves."

But our framers, who hoped that the House would be the "immediate representatives of the people . . . themselves," did not take into account the artifice, sleight of hand, or cynicism of modern politicians. Because of state-by-state deals between the political parties, the House of Representatives has become almost impervious to popular influence, its makeup long predetermined by the party leaders, its members a self-perpetuating oligarchy that listens to no one but their own political bosses.

The Boss Is Back

The reapportionment of 2001–2002 was different from any that came before. Politicians were aided in their gerrymandering duties by sophisticated new computers and software not available in previous years. Columnist Bill Kristol cites this cyber-revolution as a key element in the successful gerrymandering of 2002. But as they say on the network news, there was a human face to this tragedy, too: the cigar-chomping scowl of the old-fashioned party boss. After years of retreat in the face of a politics dominated by television advertising, the boss of old reemerged as the unsung hero of the House elections of 2002. Those he favored, won. Those he disdained, lost. It was as simple as that.

It's ironic: With all the well-meaning focus and fuss about money as a corrupting factor in politics, most commentators have missed the role of reapportionment—a silent redrawing of the American political landscape that threatens to leave one half of our legislature frozen in time for the next decade. Let the McCains and the Feingolds pursue campaign finance reform. Let the editorial pages rant about elections being bought and sold. Here, in their back rooms with their trusty laptops, the party leaders held sway, impervious alike to the will of the voters and the demands of good government. They were back in business.

Political bosses have had a tough life in the past half century. After dominating both the nomination and election process for much of the nation's history, the leaders lost most of their power through reforms and the growth of television.

Party reforms enacted in the early 1970s required that nominations for Congress and other offices be decided largely through primaries, rather than by bosses. No longer did aspirants have to grovel before their local party leaders for the nomination. Now the Democratic and Republican candidates were decided in primary contests by the voters themselves.

Television completed the disempowerment of party bosses. With TV airwaves accessible to anybody with money, the ward leader was suddenly displaced by the networks as the most important factor in winning votes. The door-to-door work of party foot soldiers was replaced by mass advertising, and the advantage suddenly shifted to the candidates best loved not by their party leaders but by their local bankers.

To adjust to this loss of esteem and clout, party leaders scrambled to control the fund-raising machinery of the political process. By setting up campaign committees at the national and state levels, they sought to tie up most of the major campaign contributors so that they would rain favor and checks only on those anointed by the party hierarchy. But it never seemed to work that way. Rich candidates were always spoiling their calculations, and even some less wealthy men and women proved excellent fund-raisers, quite capable of earning their own campaign funds.

But the party leaders had one ace in the hole: reapportionment. Their control over the redistricting process is absolute. And with this comes the ability to punish and reward congressmen, a power the bosses of old would have envied. When a congressman runs afoul of his party chieftains, he can expect retribution on Election Day. The growing power of state party leaders, through their ability to control reapportionment, is matched by the growing power of legislative leaders in Washington. In the recent redrawing of district lines, those who regularly fell in line and backed the party were rewarded with good districts, while the others had to fend for themselves.

And the key to the changes of 2001–2002, of course, was the census of 2000. The way the district lines were drawn in the reapportionment that followed not only affected who would win in November 2002 but set the pattern for who would control the House for the next decade.

Usually, reapportionment triggers a food fight that brings out the worst in our politicians. When states gain or lose seats in the 435-member House, a game of political musical chairs ensues in which congressmen compete to protect their seats from changing when the music stops and the new lines are drawn.

The new census numbers of 2000 revealed major shifts in population, moving congressional seats from states that had lost people (usually in the Northeast) to Sunbelt states that had gained population.

Twelve seats switched from states that had lost relative population to states that had grown faster. In all, eight states gained seats and ten lost them.

The gainers were:

Florida (+2)	California (+1)
Georgia (+2)	Colorado (+1)
Texas (+2)	Nevada (+1)
Arizona (+2)	North Carolina (+1)

And the losers:

Connecticut (-1)	Ohio (-1)
Indiana (-1)	Wisconsin (-1)
Illinois (-1)	Oklahoma (-1)
Michigan (-1)	New York (-2)
Mississippi (-1)	Pennsylvania (-2)

Democrats knew that when states like New York, Pennsylvania, Michigan, Illinois, Connecticut, and Wisconsin lost seats, their party was likely to lose its total number of members in Congress. They also knew that most of the new seats in Florida, North Carolina, Texas, and Arizona were likely to go Republican. After all, these states all went for Bush (or so we're told).

So when it came time to start redrawing the district lines, the Democrats approached the Republicans and came to a tacit understanding: The Republicans could improve their chances of picking up most of the new seats and the swing districts where there were vacancies—as long as they allowed the Democrats to increase the Democratic vote in the seats held by their must vulnerable incumbents. That way, incumbents of both parties would be virtually guaranteed reelection for the whole decade, even if it gave the GOP an edge in the House for the foreseeable future.

Congressional Quarterly commented on the deal, noting, "In most . . . states, Republicans were able to broker incumbent protection plans with Democrats that sharply limited the number of competitive districts in play: Only 45 seats were ranked by *Congressional Quarterly* as highly competitive just before the election."

Senate candidates may have had special-interest donors to thank for their victories in 2002, but House members had to call their local party leader, the state legislature's boss, and their Washington party apparatchiks to genuflect humbly for their good fortune on Election Day.

Money Didn't Matter—District Lines Did

Elections for the Senate are fought with money. But given the decisive role of redistricting, when it comes to House elections money doesn't have nearly as much to do with it. As long as your district is drawn to your advantage, it doesn't matter so much whether you spend millions or only hundreds of thousands of dollars on ads. It isn't that important how hard you campaign or even whether you're serving particularly well. It's the district lines that determine whether or not you get to stay in the House.

In fourteen of the most competitive House races of 2002, the winners spent a total of $25 million; the losers actually shelled out *more*— $27.7 million! In those fourteen races, the winner substantially outspent the loser in only five contests. The loser spent a lot more than the winner in four races, and the winner and the loser were within 20 percent of each other in spending in the remaining five districts.

The House elections of 2002 were decided by gerrymandering. While columnists and critics, reporters and reformers focused on the pernicious influence of money in politics and touted legislative and administrative remedies to redress it, the political regulars were chuckling to themselves as they determined the outcome of the House elections with computers, red pencils, and maps.

The Hill, a weekly magazine published about Congress, describes the process as political "insider trading, just like Enron or Martha Stewart—except this is political."

The Great American Sinecure

The real injustice of the reapportionment deal of 2001–2002 was that it created a kind of permanent membership for the House of Representatives—a body of incumbents whose jobs would be ensured for at least another decade.

In 2002, only sixteen incumbent representatives were defeated for reelection, an incredibly low total. By contrast, in 1992, after the reapportion of 1990, forty-three House incumbents lost their seats— almost three times as many. In 1982, after the census of 1980, thirty-nine incumbents were defeated.

And, of the sixteen House incumbents defeated in 2002, eight lost *not to a challenger* but to a fellow incumbent thrown into the same district by reapportionment. In four others, the incumbent was defeated by a nonincumbent challenger in a primary. In the entire United States, *only four incumbents* lost their seats to nonincumbent challengers of the opposite party!

Not only did virtually every House incumbent win in 2002, but very few of the races were even close. In only forty districts (less than one-tenth of the total) was the winner elected with less than 55 percent of the vote. In the other 401 races, there was no competition to speak of.

Thirty-three House members had no opposition at all in the general election, and another eighty-two might as well have been unopposed, winning with more than three-quarters of the vote. In all, 62 percent of the congressmen and -women elected in 2002 won their seats with 65 percent of the vote or more. This chart shows how few of the races were genuinely competitive:

WINNING VOTE SHARE IN 2002 HOUSE RACES

Vote Share of Winning Candidate	Number of Seats
Less than 55%	40
55% to 64.9%	112
65% to 74.9%	168
75% or more or unopposed	115

The most dramatic way to appreciate the impact of reapportionment on the 2002 election, though, is by looking at the success that year of congressmen who won only narrowly in 2000. These congressmen, the swing-seat winners of 2000, were lifted out of harm's way by the reapportionment and the blessings of their party leaders.

Eighteen of them—almost half—were reelected with more than 60 percent of the vote. Nine others got between 55 and 60 percent of the vote. All but seven saw their margin of victory increase after their districts were redrawn.

Reapportionment was the key factor in these incumbent victories. When the Center for Voting and Democracy analyzed the fates of House members who won close races in the 2000 election, it found that through reapportionment three-quarters of the incumbents "had their districts made safer in that the district was redrawn so that the presidential candidate of their party won a higher percentage of the vote in their [new] district."

The House has been leaning further and further toward oligarchy over the years. Fewer and fewer candidates are elected in genuinely competitive districts. The following chart compares the number of

House members who were elected by less than 60 percent of the vote over the past five election years:

NUMBER OF HOUSE SEATS WON WITH LESS THAN 60% OF THE VOTE

Year	Number of Seats Won with Less Than 60% of the Vote
1994	168
1996	177
1998	119
2000	130
2002	99

If this keeps up, our House races are going to start looking like Election Day in Saddam Hussein's Iraq: popularity contests where the winning candidate takes home 100 percent of the vote . . . largely because he was running unopposed.

The Impact of a Noncompetitive Congress

The competitiveness that kept Congress on its toes in the early 1990s, then, has given way under Republican domination to the kind of self-perpetuating aristocracy the framers thought the Senate would become.

Indeed, the *Cook Political Report* (which is to elections what the racing tout sheet is to the track) ranked only 55 House races nationally as "competitive" in its preelection listings, compared with 121 in 1992 after the last reapportionment.

More ominously, Charlie Cook, the report's author, feels that the trend toward noncompetitive districts will continue and even accelerate. "Perhaps most alarming about this decline in competition is that, typically, greater competition and turnover characterize the first couple of congressional elections after redistricting. The legislators settled into their new districts and the level of competition goes down until new maps are drawn. If the competition is this low in the first election after a redistricting, imagine what it will be like in 2008 or 2010."

According to *The Hill*, the result of the increasing number of safe seats for incumbents has been that "most voters have become bunkered down in safe, one-party districts where their only viable choice is to ratify the candidate—usually the incumbent—of the party that dominates their district. If you are a Democrat in a solidly Republican district, a Republican in a solidly Democratic district, or a supporter of a minor party, you don't have a chance of electing your candidate, no matter how much money your candidate spends."

The impact of one-party districts on voter turnout is evident. Nationally, only 39 percent of America's voting age population went to the polls in the 2002 midterm election, because there was no contest in most districts. But where there was a fight, voters voted. For example, in South Dakota, where Democratic senator Tim Johnson had the fight of his life in beating back Congressman John Thune, the Republican challenger, and in the Minnesota Senate race, where St. Paul mayor Norm Coleman defeated the old Democratic warhorse Walter Mondale after Paul Wellstone's death, voter turnout soared to 61 percent of voting-age adults.

Just as special interests benefit from the flood of money into political campaigns, so the party bosses reap the rewards of the pivotal manipulation of reapportionment to protect incumbents. Their message to those who finally sit in the halls of the Congress is simple: "We made you . . . and we can unmake you, the next time the district lines are drawn."

The result is a rapid increase in party solidarity and in the tendency for congressmen to vote as a solid phalanx just the way their leaders tell them to. When one's electoral fate rests not on the ability to campaign, or on constituent service, or even on the amount of campaign cash one can raise, but on the favor of the party bosses in Congress and in the state legislature, a congressman becomes a puppet rather than an independent-minded legislator. Increasingly, this tendency brings American politics in line with the British system, which led Gilbert and Sullivan to opine:

> I always voted at my party's call
> and I never thought of thinking for myself at all.
> I thought so little, they rewarded me
> by making me the ruler of the Queen's Navee.

And the comparison with Britain doesn't end with party regularity in voting. During the nineteenth century, the bane of British democracy were the "rotten boroughs" or "pocket boroughs" that dominated the House of Commons. These seats typically represented districts that were once populated but now consisted only of a manor house, its lord, and a staff of servants. So small were the populations of these districts that a handful of aristocrats could regularly outvote the mass of urban voters who elected members of Parliament from the big cities.

Democracy is on a similar trial in the United States as a result of gerrymandering. While all congressional districts are equal in population—we learned that much from Britain's trials and difficulties—the grouping of voters into single-party districts effectively disenfranchises most Americans, as surely as the rotten boroughs did in England.

Because of the default of democracy in the House of Representatives, the U.S. Senate has become the chamber that best reflects the movement of opinion in the American democracy. But since only one-third of the Senate is elected every two years, the impact of public opinion is by definition moderated and slow to register. It takes six years to change the complexion of the Senate. Only the House can change completely every two years.

The impact of the erosion of democracy in the House of Representatives not only effects voters, it inculcates a culture of arrogance in the members. How responsive will a House member be when he or she is elected by two-thirds or three-quarters of the vote? Can a body with so little turnover and so few competitive seats be called truly democratic?

David J. Garrow, Emory University Law School professor and Pulitzer Prize–winning historian, summed up the 2002 election results aptly in a *New York Times* op-ed article. "Judging from last week's elections," he writes, "the House . . . has become uncompetitive, sclerotic, and immune to change. The culprit is gerrymandering of Congressional districts. If reform is not enacted soon, democratic choice will be sapped out of the House altogether."

Iowa Offers the Solution

What's the answer? How can we restore democracy in the House of Representatives?

Tiny Iowa provides a glimpse of the reforms we must adopt. While only 10 percent of the 435 seats elected to the House nationwide were competitive in the 2002 elections, 4 of Iowa's 5 seats were closely fought. With just 1 percent of America's population, Iowa had one-tenth of the competitive seats in the House elections of 2002.

Why? Because in 1981 the Iowa State Legislature and the governor agreed to delegate reapportionment to a nonpartisan Legislative Services Bureau, which drew the lines subject to legislative approval and gubernatorial veto. Under Iowa law, the bureau must consider "four, essential, measurable criteria" in drawing the lines: "Population equality, contiguity, unity of counties and cities . . . and compactness." It cannot consider "political affiliation, previous election results, the addresses of incumbents, or any demographic information other than population." While the Republican-controlled legislature had to approve the lines that eventuated, it was allowed only a straight vote, up or down, with no amendments.

"We don't even look at the current districts," said Ed Cook, legal counsel for the Legislative Service Bureau that redraws the districts. "We just start over."

The politicians have accepted the redistricting process philosophically. "We're just all calmly sitting back, waiting for the first plan, and then making sure we're prepared to analyze it when it comes out," says Marlys Popma, the executive director for the Republican Party of Iowa. "If you're looking for fair, I think the Iowa system is best."

"It cuts out all of the partisan rhetoric," observes Sarah Leonard, a spokeswoman for the Iowa Democratic Party.

The *Economist* has also pointed out that "in one respect, Iowa towers above the rest of America like a silo above the cornfields: democracy. . . . No method of redistricting is perfect, but the Iowan way is better than the rest. Forget the jokes about silos. Maybe it is time to tow the Statue of Liberty to Des Moines."

All four congressional incumbents in Iowa were reelected, just as many House incumbents might be in fairly drawn districts. But they each had a tough fight—invigorating for democracy and important in keeping them in touch with their districts. Republican incumbents Congressman Jim Leach won with 52 percent of the vote, Congressman Tom Latham got 55 percent, and Congressman Jim Nussle won with 57 percent. The state's lone Democratic incumbent, Leonard Boswell, was reelected with 53 percent of the vote.

Voters in Arizona recently took matters into their own hands and voted, by referendum, to adopt a system similar to that in Iowa. Until voters in the other states that permit referenda and voter initiative make a similar move, we will be effectively disenfranchised in electing the House of Representatives.

What Can We Do About It?

If you live in one of the states that have laws that allow referenda and initiatives—about half of them—there's plenty you can do. Organize an effort to save democracy by taking reapportionment out of the hands of the politicians.

Collect the signatures! Put an initiative on your state ballot to:

- Undo the reapportionment of congressional districts in your state and require a new set of lines by the election of 2006 (your initiative will be on the 2004 ballot).
- Require that the reapportionment be handled by a nonpartisan bureau, as in Iowa, and that the politicians have no say in the outcome.
- Forbid the bureau from considering incumbency, party, or voting patterns in the districting.
- Require that the districts be compact and contiguous.

Once you put it on the ballot by getting enough signatures, don't worry, it will pass. The state's newspapers will support it, and the public will see the justice in it. And the politicians won't be able to do a thing about it. All their money? All their power? It's all meaningless in the face of the people's will. As the nursery rhyme says, "*All the King's horses and all the King's men couldn't put Humpty Dumpty together again!*"

Whether it's through the Iowa solution or some other approach, the bottom line is this: We need an antitrust law for our politicians.

In the private sector, we invite businessmen to compete—the tougher the better—in the sure knowledge that their products will grow better and cheaper if they do. But when the businessmen get together, in "restraint of trade," and carve up the world between them, we sue them under our antitrust laws and break up the combination.

In effect, we must now do the same thing with our politicians—break up the conspiracy to fix elections through the unfair reapportionment of voting districts. You can't trust politicians not to take as much political advantage as they can, even if that means cutting a deal with the other side.

Just as we took our marketplace back from monopolies and trusts of the robber-baron days through antitrust laws, we need to take our democracy back from the coalition of the parties.

Benjamin Franklin said the Constitution gave us a democracy—"if you can keep it." Now, we face one of the tests he must have had in mind.

EIGHT

THE ATTACK ON OUR KIDS:

HOW THE GOVERNORS SWIPED THE ANTITOBACCO MONEY AND ENDANGERED OUR HEALTH

*A*UTHOR'S WARNING: *My mother died of tobacco-induced cancer, as did my father-in-law. I worked with the attorney generals in helping to formulate their lawsuit against tobacco companies and to persuade President Clinton to order the FDA to oversee tobacco advertising. So I am not impartial when it comes to smoking. But I am passionate.*

Is there anything worse one can say about a leader than that he had the chance to save lives and didn't? That's a fact Governor Gray Davis will have to live with for the rest of his life. The California governor has taken a state that was leading America in preventing smoking deaths, and singlehandedly reversed its progress, paving the way for the tobacco industry to indoctrinate another generation of millions of young California smokers.

The 1990s saw state attorney generals like Mississippi's Mike Moore win the greatest advance in public health of the past fifty years. But now governors like Davis are giving it all away.

And our children will pay the price, to the tune of decades of premature death and billions in health-care costs.

Moore and his colleagues sued the big tobacco companies at the start of the 1990s. Because of their guts and vision, Big Tobacco had to agree to pay $240 billion over twenty years—including almost $9 billion this year—to the states, to compensate them for the cost of treating those whose lives tobacco had ruined. The idea was to spend the money persuading kids not to smoke.

But then Davis stepped in. Spurred by a huge budget deficit, he and his fellow governors raided the tobacco settlement—and diverted the money that could have paid for anti-smoking campaigns to fund their regular operating expenses. Davis even committed the money for years into the future, to back up bonds he issued to pay for his big spending.

OFF WITH HIS HEAD!

What's at Stake?

The stakes could not be higher. Tobacco causes approximately 430,000 deaths annually, about one-fifth of all fatalities in the United States. It dwarfs other causes of death, which get a lot more attention in the public media. Homicides? Eighteen thousand deaths each year. AIDS? Another eighteen thousand. Car accidents? Forty-five thousand. Suicides? Thirty thousand. Smoking? Nearly half a million lives!

Among cancers, lung cancer towers over all others as a killer, destroying 155,000 lives each year—more than colon, prostate, and breast cancers combined. And smoking causes 87 percent of all lung cancers.

When pregnant women smoke, the results are particularly disastrous. According to the American Lung Association, smoking in pregnancy accounts for about a quarter of all low-birth-weight children, one in seven postterm deliveries, and one in ten of all infant deaths. The association warns: "Even apparently healthy, full-term babies of smokers have been born with curtailed lung function."

I have to admit that part of my anger about smoking stems from my own very premature birth, to a mother who was a heavy smoker. Before anyone (except perhaps the tobacco companies) knew that smoking caused problems for fetuses, I was born, at two pounds eleven ounces, in 1947. I survived only by good luck and great care. My mother died of colon cancer in 1993, after a heart attack had almost killed her. In the recovery room, she was still smoking.

Tobacco deaths aren't cheap, and they certainly aren't quick or painless. On average, smokers have $12,000 more in lifetime medical costs than nonsmokers do. Federal, state, and local governments have to spend over $50 billion a year to treat smoking-caused diseases, $166 for each man, woman, and child in America.

And yet, at least four thousand kids light up for the first time each and every day. One in three will die from the habit. Even today, one twelfth grader in three smokes.

The Great Breakthrough:
The Attorney Generals Beat the Tobacco Companies

In the last twelve years, however, a few principled souls had found a way to turn this trend around, to bring tobacco to its knees and free our children from the menace of an early grave.

It all began in 1991, when three Mississippi lawyers, all deeply spiritual men, decided that enough people had died from smoking and that it was time to stop. Mike Moore, the state's attorney general, along with anti-asbestos lawyers Dick Scruggs and Steve Bozeman, hatched what others called a harebrained scheme to sue the tobacco companies for the money Mississippi had to spend treating people with smoking-related lung cancer, heart disease, or emphysema. After all, they reasoned, why should the taxpayers have to pick up the tab when the tobacco companies had gotten all the profits?

Lots of people had sued the tobacco companies, but nobody had ever won. Juries kept asking a simple question: Who's making these people smoke? Aren't they lighting up of their own free will?

But Moore, Scruggs, and Bozeman had come up with a new approach. They asked a new question of their own: What about kids who start smoking for the first time? At twelve and thirteen years old, are children really able to make informed decisions about their health, lives, and future? Cigarette companies target kids in their ads, showing how cool Joe Camel is or how sexy women who smoke Virginia Slims can be. By exploiting the gullibility and insecurity of early teenagers, the tobacco companies get them to try cigarettes. Like drug pushers, the tobacco companies even offered free minipacks of cigarettes to young people all over the world, to induce them to start, according to the Campaign for Tobacco-Free Kids (CTFK).

As soon as young people start smoking, of course, chemistry takes over. The nicotine induces a craving, which makes the kids keep coming back for more until they become fully addicted, several-packs-per-day smokers. By targeting kids, the tobacco companies were playing unfair.

These adults never made a considered, mature decision to smoke. They'd been hooked as kids, and by the time they were adults they were addicts.

Even *smokers* agreed that kids shouldn't start smoking. Most smokers remembered that they had started as young adolescents, and they knew all too well how hard the habit was to shake—even if your life depended on it—once tobacco had sunk its claws into your lungs.

By focusing on kids, the trio had come up with a winning antismoking strategy. So, led by Mike Moore, they decided to aim their attack at teen smoking.

Scruggs and Bozeman dug up evidence that the tobacco companies had manipulated their cigarettes to make them more addictive. They proved that the companies were concealing data that showed how unhealthy tobacco is, that they had lied when they said their product did not cause cancer, and that they deliberately targeted kids in their promotions and ads.

Moore demanded that the tobacco companies rein in their shameless promotions aimed at hooking kids. No more billboards near schools, ads in magazines kids read, displays of cartoon characters to depict smoking as cool, or giveaways of free packs in areas frequented by young people.

But that wasn't all. Moore also came up with another brilliant idea: Get the states to run ads on television to persuade kids *not* to smoke— and get the tobacco companies to pay for them as part of the verdict in their lawsuit!

"Turn off one spigot and turn on the other" was the way Moore described his strategy. "Stop the ads and marketing gimmicks that are inducing kids to smoke and fund antitobacco ads aimed at children to get them to stop smoking or never to start."

The attorney generals did their part. Led by good men and women like Mississippi's Moore, Florida's Bob Butterworth, Connecticut's Richard Blumenthal, and Louisiana's Richard Ieyoub, they each sued the giant tobacco firms, in separate suits, demanding that they reimburse the states for the hundreds of billions of dollars they had to pay to treat sick smokers. Because the states had no choice but to treat patients who were eligible for Medicaid, the attorney generals argued, Big Tobacco should bear its fair share of the cost.

The battle to win the lawsuit was arduous. Few felt that the effort would succeed. But then President Bill Clinton entered the battle.

For all of his failures in the war on terror, Bill Clinton proved decisive in winning the war on smoking. He never stood taller than he did in August 1995, when he announced he was approving regulations that would allow the Food and Drug Administration to ban cigarettes ads aimed at children, and empowering the agency to label tobacco as an addictive drug and cigarettes as a drug-delivery device (syringes for the lungs!).

Erskine Bowles, the president's friend and future chief of staff, warned Clinton about tangling with the tobacco companies. "I just don't like the idea of fighting someone who has a billion dollars to spend on advertising against us," he warned as the president pondered whether to weigh in against the cigarette companies. But Vice President Al Gore brought Clinton around when he told the story of how he'd fought for warning labels on cigarettes—and survived politically in Tennessee despite it. "Smokers, even tobacco farmers, don't want kids to smoke," the vice president said.

Clinton's courageous stand may have cost him North Carolina, and nearly led to losing Tennessee and Kentucky, when he ran again in 1996. But it animated the antismoking crusade as nothing else could. Under attack from the most prominent pulpit in the land, Big Tobacco began to wither and run for cover.

Besieged by the growing chorus of antismoking opinion, and labeled virtual drug dealers by the president, tobacco executives sought out the attorney generals to find out what peace terms they could get.

The answer came back quickly: pay $240 billion over twenty years, implement the restrictions on advertising and marketing to teens, and we'll drop the lawsuit. Former Surgeon General C. Everett Koop, who had done so much to fight smoking, and FDA administrator David Kessler, an antismoking zealot, called on the attorney generals to stand firm in their demands and turn down the tobacco companies' offer.

But eventually, the attorney generals and the tobacco companies settled. The cigarette companies, in effect, folded and knuckled under. One spigot was turned off—the ads and marketing to kids were gone (au revoir and good riddance, Mr. Joe Camel)—and the other turned on: $240 billion flowing to states to fight against teen smoking.

But there was a catch.

The $240 billion had to pass through the governors, and their state legislatures, before it could make it to the front lines of the antismoking battle. This was a huge cash influx, and from the start the governors must have been tempted to divert the money to other, more selfish purposes. But times were good in the late 1990s, and state revenues abundant, so most governors stood up to the pressure, allocating large amounts to smoking prevention programs. Not until the recession of 2001–2002 tightened its hold on the economy would things start to change for the worse.

How States Can Cut Smoking

Moore knew one thing for certain: If the states got the money to run ads against smoking, they could win the battle. He had only to look at what had already happened in California.

California voters had taken matters into their own hands in 1988, when they approved Proposition 99, the California Tobacco Tax Initiative. The ballot measure, which passed overwhelmingly, increased cigarette taxes by 25 cents per pack, with 20 cents dedicated to tobacco prevention and control and 5 cents to tobacco-related disease research. As a result of the voters' wisdom, for more than a decade California spent $85–$115 million per year trying to persuade people not to smoke.

The results have been astonishing. In California, lung cancer rates have dropped 14 percent since 1988, whereas national lung cancer rates fell by only 3 percent. Because of its farsighted efforts, more than two thousand lives are saved each year in California due to the drop in lung cancer rates alone.

And this savings doesn't count reductions in deaths from heart disease and emphysema. Nor does it really count the impact of youth antismoking efforts. Kids, unlike older adults, don't die of smoking until later in their lives, so the impact of current antiteen smoking campaigns won't kick in for several decades.

Since 1988, Proposition 99 has brought California dramatic results:

- Cigarette consumption fell 58 percent, twice as fast as in the nation as a whole.
- Smoking among twelve- to seventeen-year-olds dropped by 35 percent.

- The proportion of adults who smoked dropped from 23 percent to 17 percent—twice the rate of decrease as during the previous decade.
- More than 1.3 million Californians have quit smoking.
- Half as many people are exposed to indoor secondhand smoke as before Proposition 99.

Can state governments stop kids from smoking? You bet they can.

Massachusetts began spending money to cut teen smoking in the early 1990s under the administration of Governor William Weld, a progressive Republican. As the National Institute of Health reported: "A recent evaluation of the Massachusetts tobacco control program showed a 15% decline" in smoking, "compared to very little change nationally—thus reducing the number of smokers there by 153,000 between 1993 and 1999."

In Florida, a school-focused program backed up by advertising was responsible for cutting smoking among middle-school students almost in half and, in high schools, by roughly one-third.

In Minnesota, Target Market, an antismoking group, ran a campaign called "You Target Us, We Target You." Writing for *Consumer Health Interactive,* Loren Stein describes how they did it: "With an $8.5 million yearly budget:"

- Teen volunteers "produce concerts and sports events to publicize the cause."
- They "hold 'Kick Ash Bash' gatherings and sell music CDs, underground films, and CD-Roms."
- "They sponsor anti-industry presentations, including 'tours' of the tobacco industry's most damning documents, and campaigns such as 'Rip It Out,' in which teens tear out tobacco magazine ads and mail them to Target Market in order to receive the organization's free promotional gear."
- "Working as advisers to ad agencies, teenagers in Minnesota brainstorm ad concepts for a statewide media blitz that highlights tobacco industry manipulation of young people. One award-winning television ad showed an African American teen reading from tobacco industry documents, including the quote, 'We target the young, the black, and the stupid.'"

Oregon achieved similar results. Maine cut high-school student smoking by one-third, and in Mississippi, where all the antismoking fuss started, it dropped by a quarter.

The antismoking ads are fierce. Those who live by the sword, die by the sword! The same tool Big Tobacco uses to hook kids on smoking—advertising—turns out to be the most effective weapon in stopping them from lighting up.

One ad features a coffin, with the label CRUSH PROOF BOX. Another shows Joe Camel hooked up to an IV. In another, a paunchy, middle-aged, balding man sits in a bikini on a beach smoking a cigarette, with the caption "no wonder tobacco executives hide behind sexy models." In a fourth ad, Patrick Reynolds, disinherited heir to the tobacco family, talks about the effects of smoking, and the duplicity of his family company.

One particularly memorable ad shows childhood photos of ex-smoker Pam Laffin, who says:

> I started smoking when I was ten because I wanted to look older. And I got hooked. Cigarettes gave me asthma and bronchitis, but I couldn't quit. I didn't quit until I got emphysema and had a lung removed. I was twenty-four. I'm twenty-six now. My medication, which I'll take for the rest of my life, left me with this fat face and a lump on my neck. I started smoking to look older, and I'm sorry to say . . . it worked.

Tough and effective.

The Governors Sell Out

But then, in 2000, the economy began to go south. State revenues fell and deficits loomed ahead. And, predictably, governors and state legislators started eyeing the tobacco settlement money as a quick and easy way to paper over their looming budget deficits. No doubt hoping that the voters wouldn't notice, they applied the money elsewhere before it could serve the purpose that the crusaders had intended—to stop kids from getting a deadly habit. Led by Gray Davis in California, the governors used their budget-trimming scissors to knife this revolutionary antismoking ad campaign in the back, ducking the consequences of their own irresponsible overspending in the process.

Like the political coward that he is, Davis decided to issue bonds to borrow his way out of his budget deficit. Not surprisingly, Wall Street demanded evidence that California would be able to pay the debt service on these bonds. Since the borrowing isn't designed to float bonds for any revenue producing activity (like a toll road, for example), potential bond buyers insist on collateral before they'll lend California the money to pay its bills.

So Davis pledged the tobacco settlement money for the next ten years ($4.5 billion worth) to repay this year's borrowing. Then, adding insult to injury, he also proposed to cut funding in half for tobacco-prevention activities. Not only are Davis and the legislature condemning thousands of Californians by cutting back antismoking efforts this year, but he is destroying the state's ability to fight smoking in future years, by diverting the tobacco settlement money en masse so he won't have to raise taxes to pay for his big spending habits.

Then, to top it off, Davis and the legislature refused to raise cigarette taxes—the one move that would have helped both to bridge the deficit and to save lives by cutting tobacco consumption.

And Davis wasn't the only one who gave in to temptation. In Arizona, Governor Jane Dee Hull has also cut the antismoking program in half, overriding a 1994 vote of the people to raise the state sales tax on tobacco and use the money to fund smoking prevention. Not only does Governor Hull not care about the health of her citizens, but apparently she doesn't care much about their political mandates either. Fortunately, in November 2002 Arizona voters took matters into their own hands and passed Proposition 303, which raised cigarette taxes and mandated that much of the resulting revenues go to antismoking efforts. Still, the politicians may yet try to get their hands on the money first.

Governor George Ryan of Illinois made a national reputation for himself by sparing death row inmates in a massive commutation of capital sentences before he left office at the end of 2002. Unfortunately, his desire to save lives doesn't seem to have extended to the noncriminal elements in his state as well. Last year, Governor Ryan slashed tobacco prevention funding by 74 percent, from approximately $46 million to only $12 million. Even after raising the cigarette tax by 40 cents per pack, he refused to devote a dime of the extra money to stopping smoking among the state's teens or adults.

But the picture isn't entirely bleak:

- Under Governor Angus King, Maine has increased antitobacco funding.
- Governor Jim McGreevey of New Jersey saw to it that when his state raised cigarette taxes 70 cents per pack, the money would go to saving lives. Of the seventeen states that raised cigarette taxes this year, New Jersey is the only one to commit part of the money to antismoking efforts.
- Indiana, Maryland, Alaska, Minnesota, Pennsylvania, Missouri, Arkansas, and Washington State maintained or increased their large investment in cutting smoking.

But there's much more work to be done. Using an A-to-F grading system, the American Lung Association has ranked the fifty states on their performance in funding antitobacco efforts. Only nine states got an A or B; thirty-two others, plus the District of Columbia, flunked with an F. Here's the list:

HOW STATES MEASURE UP IN ANTISMOKING EFFORTS

Grade	States (total)
A	Arkansas, Indiana, Maine, Maryland, Minnesota, Mississippi (6)
B	Pennsylvania, Vermont, Washington State (3)
C	Alaska (1)
D	Arizona, California, Colorado, Delaware, Hawaii, New Jersey, Virginia, Wyoming (8)
F	Alabama, Connecticut, District of Columbia, Florida, Georgia, Idaho, Illinois, Iowa, Kansas, Kentucky, Louisiana, Massachusetts, Michigan, Missouri, Montana, Nebraska, Nevada, New Hampshire, New Mexico, New York, North Carolina, North Dakota, Ohio, Oklahoma, Oregon, Rhode Island, South Carolina, South Dakota, Tennessee, Texas, Utah, West Virginia, Wisconsin (33)

It's time to face down those politicians whose states don't measure up, and make them take responsibility for their actions, political responsibility, in the form of defecting voters and lost elections. But, first, here are the nine states and their governors who deserve credit—and the thanks of their constituents—for getting an A or a B. These states are spending at or close to the levels the Centers for Disease Control recommends to save the lives of smokers and potential tobacco users:

THE ANTISMOKING HONOR ROLL:
STATES THAT SPEND ADEQUATELY ON TOBACCO PREVENTION AND CONTROL

Indiana	Governor Frank O'Bannon (Dem)
Maine	Governor John Baldacci (Dem)
Maryland	Governor Robert L. Ehrlich Jr. (Rep)
New Jersey	Governor Jim McGreevey (Dem)
Minnesota	Governor Jesse Ventura (Ind)*
Mississippi	Governor Ronnie Musgrove (Dem)
Arkansas	Governor Mike Huckabee (Rep)
Pennsylvania	Governor Tom Ridge (Rep)*
Washington	Governor Gary Locke (Dem)

*Former governors who nevertheless deserve credit for progress made during their terms.

But how about the states that flunked? What damage did these governors cause? Here's how to quantify it:

In its fourteen years of antismoking efforts, California cut lung cancer deaths by 14 percent. Had the cancer rates dropped all across the nation, 18,000 fewer people would have died. (That's about as many as would be saved if we completely eliminated all homicides in the United States for a year.) So let's see: how many people did each of the *flunking* governors allow to die by not adopting similar measures?

NUMBER OF LIVES EACH STATE COULD HAVE SAVED
BY REPLICATING CALIFORNIA'S ANTISMOKING EFFORTS

State	Governor	Lives That Could Have Been Saved
Alabama	Bob Riley	284
Connecticut	John G. Rowland	218
Florida	Jeb Bush	1,012
Georgia	Sonny Perdue	523
Idaho	Dirk Kempthorne	83
Illinois	Rod R. Blagojevich	793
Iowa	Tom Vilsack	187
Kansas	Kathleen Sebelius	172
Kentucky	Paul E. Patton	259
Louisiana	Mike Foster	286
Massachusetts	Mitt Romney	407
Michigan	Jennifer Granholm	635
Missouri	Bob Holden	358
Montana	Judy Martz	58
Nebraska	Mike Johanns	110
Nevada	Kenny Guinn	128
New Hampshire	Craig Benson	79
New Mexico	Bill Richardson	117
New York	George E. Pataki	1,213
North Carolina	Mike Easley	514
North Dakota	John Hoeven	41
Ohio	Bob Taft	725
Oklahoma	Brad Henry	221
Oregon	Ted Kulongoski	220
Rhode Island	Don Carcieri	67
South Carolina	Mark Sanford	257
South Dakota	Mike Rounds	49
Tennessee	Phil Bredesen	364
Texas	Rick Perry	1,334
Utah	Michael Leavitt	142
West Virginia	Bob Wise	115
Wisconsin	Jim Doyle	343

Is this a brutal way to look at the question? Maybe. Is it justified? You bet. Remember, these governors put the needs of their own political convenience ahead of the needs of their people. To every one of them, I say: *You were elected to serve and protect the people of your state. But you've shown you care more about preserving your budget than about their health.*

OFF WITH YOUR HEADS!

When the governors decided to cut antismoking programs, they weren't just being shortsighted, they were being stupid. Cutting smoking is the best way to cut state spending.

Here's how:

When fewer pregnant women smoke—or breathe secondhand smoke—the cost of childbirth drops right away. More than half of all childbirths are paid by Medicaid; and a smoking-affected birth costs an average of $1,200 more than a healthy one.

When parents don't smoke, kids are a lot less likely to have asthma or other respiratory ailments. Parental smoking costs $2.5 billion per year in medical costs for newborns, infants, and children.

Fewer cigarettes mean fewer fires. Smoking-related fires cause more than $500 million in property damage.

For every dollar spent on antismoking efforts, almost $4 are saved in health-care spending.

Tobacco Strikes Back

While the states were slashing antismoking funding, the cigarette companies used the opportunity to ratchet up their marketing and advertising campaigns. The tobacco companies increased their marketing expenditures by 42 percent last year, to an unprecedented total of $9.6 billion annually.

As the Centers for Disease Control pointed out in its 2000 report *Reducing Tobacco Use,* "Efforts to prevent the onset or continuance of tobacco use face the pervasive, countervailing influence of tobacco promotion by the tobacco industry, a promotion that takes place despite overwhelming evidence of adverse health effects from tobacco use."

The tobacco companies say they're not targeting children in their advertising. But they are lying. Why else are they advertising? To get people to switch to their brand of cigarettes? No way. Market share among the major cigarette companies has remained relatively unchanged for years, and the idea that big tobacco is advertising just to retain or enhance brand loyalty, rather than to get more people to smoke more cigarettes, is a complete fiction. They're advertising to entice kids to start smoking. After all, children under eighteen account for two-thirds of all new smokers.

But the tobacco companies aren't the only ones to blame. Our nation's top newspapers and magazines are also in on the deal. They want the revenue they get from running gaudy tobacco ads, and they've battled to let tobacco advertise all it wants—on their pages. While the tobacco settlement bars cigarette companies from advertising in magazines that target children, it still allows tobacco companies to buy the back pages of general magazines, such as *Time* or *Newsweek*, and to advertise heavily in the print media.

If there were ever a recipe for disaster and more tobacco deaths, this is it—that tobacco is increasing its ads while the states are cutting back theirs! With tobacco spending higher than ever, it's unlikely that the gains in public health and the decreasing death rates from cancer can be sustained in the near future.

The Senate Sells Out

And it isn't just the governors who are to blame: the trail of guilt for smoking-related deaths also leads to the U.S. Senate. The senators showed that they weren't any better than the governors when they refused to pass Senator John McCain's tough proposal to cut teen smoking in 1997. It wasn't that the senators didn't know how bad cigarettes are. It's that they got too much money in contributions from tobacco to care. According to former New York City public advocate Mark Green's recent book, *Selling Out*, tobacco companies gave the senators who opposed McCain's bill an average of $17,902.

In all, the tobacco companies spent over $65 million to lobby Congress to beat the McCain bill. In addition, they spent $70 million on a media campaign to oppose it. Here are the names of the forty-two senators (all but two were Republicans) who voted with Big Tobacco to let cigarettes to continue to kill our children. Shame on each of them:

Wayne Allard (R-Colo.)

John Ashcroft (R-Mo.)

Christopher Bond (R-Mo.)

Sam Brownback (R-Kans.)

Conrad Burns (R-Mont.)

Ben Campbell (R-Colo.)

Dan Coats (R-Ind.)

Thad Cochran (R-Miss.)

Paul Coverdell (R-Ga.)

Larry Craig (R-Idaho)

Pete Domenici (R-N.M.)

Michael Enzi (R-Wyo.)

Lauch Faircloth (R-N.C.)

Wendell Ford (D-Ky.)

Slade Gorton (R-Wash.)

Phil Gramm (R-Tex.)

Rod Grams (R-Minn.)

Chuck Hagel (R-Nebr.)

Orrin Hatch (R-Utah)

Jesse Helms (R-N.C.)

Kay Hutchinson (R-Tex.)

Tim Hutchinson (R-Ark.)

James Inhofe (R-Okla.)

Dirk Kempthorne (R-Idaho)

Jon Kyl (R-Ariz.)

Trent Lott (R-Miss.)

Richard Lugar (R-Ind.)

Connie Mack (R-Fla.)

Mitch McConnell (R-Ky.)

Lisa Murkowski (R-Alaska)

Don Nickles (R-Okla.)

Charles Robb (D-Va.)

Pat Roberts (R-Kans.)

Rick Santorum (R-Pa.)

Jeff Sessions (R-Ala.)

Richard Shelby (R-Ala.)

Robert Smith (R-N.H.)

Ted Stevens (R-Alaska)

Craig Thomas (R-Wyo.)

Fred Thompson (R-Tenn.)

Strom Thurmond (R-S.C.)

John Warner (R-Va.)

President George W. Bush is unlikely to emulate Bill Clinton's antismoking stands. In his 2000 campaign, he received $6.5 million in soft money and PAC contributions from tobacco companies. In fact, as this is being written, the United States is working overtime to weaken the provisions of a new global treaty designed to reduce smoking worldwide.

The Case for Death

But if there's one outrage that beats them all, it's Big Tobacco's latest attempt to defend the role of cigarettes in our society. Having lost the battle to convince America that smoking is healthy, now they're claiming it's a handy form of population control! In 1994, W. Kip Viscusi, a Duke University economist from the heart of North Carolina's tobacco country, found—based in part on data from the Rand Corporation—that when smokers die young, states get "big savings . . . from retirement benefits, especially since smoking-related disease usually kicks in as one's working years end." Viscusi calculated "total yearly permission savings" at nearly $30 billion.

Based on ghoulish calculations like these, the *New York Times* reported that tobacco companies wanted lawsuits against them to include "an accounting of offsets, which means estimating what the states would have spent treating smokers, had they lived longer." As the CTFK summarizes the argument, tobacco companies are saying, "The federal and state governments should not invest in new efforts to reduce smoking and other tobacco use because it is cheaper to let people die from smoking . . . than to pay the new costs caused by more people living longer because they quit using tobacco or never start."

This line of argument reminds one of Jonathan Swift's famous essay "A Modest Proposal." Using the Irish potato famine to satirize laissez-faire economic theory, Swift reasoned that Ireland had two problems: too few potatoes and too many children. So, he proposed, let the Irish eat their kids!

As the CTFK points out: "Put more bluntly, [tobacco companies are saying] that smokers deserve to die early and that society should not save smokers from their own decisions. But do all smokers actually deserve to die early? Does that include those who became addicted when only 12 or 13 years old, when they were too naïve or ignorant to know any better?"

The Coming Death of Big Tobacco

Despite the cowardly retreat of the nation's governors and senators, however, juries throughout America are striking at the beast, driving a stake into the long-diseased heart of Big Tobacco.

Leading the way, the courageous families of deceased smokers have begun suing cigarette companies for damages—and winning, forcing the companies to pay huge awards. The politicians may be bought off, but the juries are coming through!

- In 1999, Mayola Williams sued on behalf of her dead husband. She said he had a dying wish: "He wanted to make cigarette companies stop lying about the health problems of smokers." The jury awarded her $79.5 million in punitive damages. William A. Gaylord, a lawyer for the Williams family, said: "The problem has been that Philip Morris and other cigarette companies have never accepted an ounce of responsibility. . . . They essentially say to their very best customers that you get what's coming to you for believing us."

- In San Francisco, Patricia Henley got $51.5 million. Ms. Henley had come down with lung cancer after thirty-five years of smoking.

- In June 2001, a California jury awarded $3 billion in punitive damages to Richard Boeken. This was the largest civil judgment ever assessed against a tobacco company.

- A Kansas jury awarded $15 million in a lawsuit against R. J. Reynolds on June 21, 2002.

- In October 2002, a Los Angeles jury ordered Philip Morris to pay $28 billion in punitive damages to Betty Bullock, who was dying from lung cancer. On appeal, a judge lopped three zeros off the award, lowering the amount to $28 million, in return for Philip Morris's promise not to appeal the verdict.

Did these people "deserve" to *make* that much money? That's not the point. The fact is that Big Tobacco deserves to *lose* that much money—and a lot more.

Each major verdict against tobacco has important financial repercussions in the tobacco industry. It raises the price of cigarettes, which in turn cuts smoking and deters teenagers from lighting up. Research has shown that each time the price of cigarettes goes up, the number of teenage smokers goes way down, and the number of adult smokers drops as well. These verdicts save lives.

And recently the news has gotten even worse for Big Tobacco. Lynn French, a fifty-six-year-old flight attendant and a nonsmoker, won $5.5 million after she charged that her chronic sinus problems stemmed from more than a dozen years aboard smoky airplanes. French's case opened the door to the potential for similar claims by twenty-eight hundred other flight attendants, and millions more who have been exposed to secondhand smoke.

Even more troubling to the tobacco giants was a $145 billion award of punitive damages in the Engle case, a class-action lawsuit on behalf of all Floridians who suffer from cigarette-caused diseases. After a trial lasting two years, on July 14, 2000, a jury found that twenty diseases (including cancers of the lung and bladder, oral cancers, heart disease, and emphysema) were caused by smoking. This verdict opened the door for individual plaintiffs to sue in Florida to collect verdicts under the class-action umbrella.

The tobacco companies are appealing the Engle decision, of course, and with the current right-wing Supreme Court cracking down on punitive damage awards, they may have a chance. In the meantime, dying ex-smoker John Lukacs has already won an award of $37.5 million on June 11, 2002, under the Engle decision.

And the government is adding to the tobacco companies' woes. Before he left office, Bill Clinton ordered the Justice Department to sue tobacco companies for the amount the federal government has had to spend through Medicare on health care for smokers. Amazingly—and to his credit—President Bush has not pulled the plug on the lawsuit, or agreed to a token settlement. This aggressive suit against Big Tobacco—similar to the successful Medicaid suit brought by the state attorney generals—could be very effective in bringing cigarette manufacturers to their knees.

Tobacco companies have countered that they pay their way through cigarette and other taxes, compensating state and federal governments for the damage their products cause. But this is not statistically true. Federal and state governments spend $50 billion treating dying and sick smokers. But they get only half as much, $24 billion, in revenue from tobacco taxes. Who pays the difference? You, the taxpayer.

With the bad news pouring in, tobacco companies—particularly Philip Morris—began to run television ads extolling the virtues of their company by pointing to their charitable works, funded with money they made from selling people cancer-causing tobacco. These ads are as revolting as they are hypocritical. As Dr. David O. Lewis, chairman of the Health Advocacy Group of Southside, Virginia, put it: "Suppose I contributed $1,000 to my church . . . and then paid $2,000 for newspaper and television ads to advertise this fact. . . . In an attempt to change their image, this is exactly what the major tobacco manufacturers are doing with their barrage of media ads touting their largess to needy organizations. They are spending more on the ads than they are contributing to the causes.

"One of their television ads concerning battered women ends with the phrase, 'after all, no one has the right to hurt you.' Give me a break. Apparently, Big Tobacco has convinced itself that giving people cancer, heart disease and strokes, and causing infants to die from Sudden Infant Death Syndrome, is not hurting these victims."

Bravo, Dr. Lewis!

The tobacco industry is bleeding. And the faster those they have sickened and killed sue—the faster this miserable industry dies—the more of us will live.

What Can We Do?

1. *Press Your Politicians to Spend the Tobacco Settlement Money on Antismoking Programs.*

Ads, antismoking treatment services, in-school programs, and other steps to stop teen smoking work well. The successful efforts in California, Massachusetts, Florida, Mississippi, Oregon, and Texas all prove that they can cut both teen and adult smoking. We need to force our politicians to make antismoking programs a top priority in allocating their budgets. Not only will full funding of antismoking programs assure that Americans live longer, but it will also reduce the cost to governments of treating sick smokers.

2. *Get the Politicians to Raise Cigarette Taxes.*

Before they raise income taxes, sales taxes, business taxes, or property taxes, make them raise cigarette taxes!

Nothing cuts smoking as successfully as higher cigarette taxes. As we've seen, tobacco companies are paying for only about half the bill we taxpayers have to foot for sick smokers. Why should we subsidize tobacco companies by paying for the damage they cause, while they pocket the profits from the cigarette sales?

As state governments consider how to raise money to bridge their increasing budget deficits, send them a message: *Raise the tobacco tax.* Not only will this tax hike raise money, it will cut state spending by reducing the number of smokers. Health experts estimate that for each 10 percent increase in tobacco taxes, teen smoking drops 7 percent and adult smoking goes down by 4 percent.

Nobody is forcing people to smoke. The tobacco tax is a voluntary tax. Why not tax cigarettes and hold down income and sales taxes?

In 2002, eleven states raised tobacco taxes:

STATES THAT RAISED TOBACCO TAXES IN 2002— AND THE AMOUNT OF THE INCREASES

Massachusetts 75 cents (per pack)	Kansas 55 cents
New Jersey 70 cents	Michigan 50 cents
Pennsylvania 69 cents	Vermont 49 cents
Oregon 68 cents	Illinois 40 cents
Connecticut 61 cents	Indiana 40 cents
Arizona 60 cents	

Massachusetts now has the highest state tobacco tax in the nation—$1.51 per pack—but New York City smokers have the highest combined state and local levy, amounting to $3 per pack. All told, it costs $7 to buy a pack of cigarettes in the Big Apple.

In states that allow referenda and voter initiatives, people should follow the example the state of Washington set in 2001, when it raised cigarette taxes by 60 cents per pack to $1.42—at the time, the highest in the nation. But be sure that the legislation requires that the extra money go to preventing smoking—so we can not only raise money, but cut costs and save lives as well.

3. *Get Your State and Local Governments to Pass Laws Restricting Indoor Smoking . . . and Get Your Boss to Make Your Workplace Smoke-free.*

Secondhand smoke contains two hundred different poisons and forty-three chemicals that cause cancer. The California Environmental Protection Agency (EPA) estimates that secondhand smoke causes the deaths of thirty-five thousand to sixty-two thousand American non-smokers from heart disease each year.

The federal EPA says secondhand smoke:

- causes one hundred fifty thousand to three hundred thousand lower respiratory tract infections in infants and children under eighteen months of age, causing seventy-five hundred to fifteen thousand hospitalizations each year
- worsens the asthma conditions of between two hundred thousand and 1 million children
- causes pneumonia, ear infections, bronchitis, coughing, wheezing, and increased mucus production in children younger than eighteen months of age

The answer? Ban smoking in as many indoor places as possible, and educate parents, particularly of young children, about the damage their habit is causing. Mayor Mike Bloomberg of New York City has taken the lead in this battle, banning smoking in all restaurants, bars, offices, and most public places. California and Massachusetts have also enacted strict anti-indoor smoke rules.

Forty-three states are rated F by the American Lung Association for their lack of efforts to curtail indoor smoking. These include:

Alabama	Hawaii
Alaska	Idaho
Arizona	Illinois
Arkansas	Indiana
Colorado	Iowa
Connecticut	Kansas
Georgia	Kentucky

Louisiana	Ohio
Massachusetts	Oklahoma
Michigan	Pennsylvania
Minnesota	Rhode Island
Mississippi	South Carolina
Missouri	South Dakota
Montana	Tennessee
Nebraska	Texas
Nevada	Utah
New Hampshire	Virginia
New Jersey	Washington
New Mexico	West Virginia
New York	Wisconsin
North Carolina	Wyoming
North Dakota	

If your state is on the list, write your governor and demand action. It is the height of injustice for a person to be made ill by somebody else's smoking.

Remember: The pernicious influence of tobacco industry campaign contributions is likely holding your state back from effective measures to guarantee your safety against indoor smoke. Only your pressure can counteract this power and make the politicians act in your interests.

4. Get Your State Governments to Restrict Youth Access to Cigarettes.

Imagine if liquor were sold in vending machines! Could states possibly enforce underage drinking laws if a kid could walk up to a machine, put in a few dollars, and get shot of booze or a beer?

But cigarette vending machines are allowed in most states, without any restriction on youth access. It's okay to have vending machines dispensing tobacco in places where kids can't go, like bars or liquor stores. But to allow them in restaurants or on street corners is practically inviting kids to smoke. Why not put them in schools while we're at it?

California practically bans tobacco vending machines, while Maine has imposed sharp restrictions on access to them. Vermont bans them entirely.

Beyond regulating or banning vending machines, states should adopt a system of issuing permits to sell tobacco—permits that can be suspended or revoked should the establishment sell to underage smokers. The system should be patterned after the alcohol regulatory system, and might even be added to the duties of that agency. Imagine how long a gas station or convenience store could stay in business if it couldn't sell cigarettes. Such regulation would be a powerful incentive to avoid sales to kids.

Fighting AIDS, crime-safety measures, antidrug programs, efforts to restrict drunk driving and enforce traffic safety, suicide-prevention efforts—these are all important. But the most powerful impact we can have on saving lives is through reducing smoking among children and adults. Those who are seriously interested in saving lives should focus on the single biggest preventable cause of death: smoking.

We don't have our national priorities straight. Think of how much time we spend worrying about homicides and crime. Consider the cost of the prison system and the criminal justice courts. We need them, of course—but remember that twenty times more people die of smoking than die from murders.

Wiping out smoking should be a major national priority. Tobacco is on the run. Lawsuits and the courage of the attorney generals has set the industry back substantially. But we have a long way to go.

I don't know what your standard is for doing good, but saving a life ranks high on my list. And there is no single act any of us can take that would do more to save lives than to fight smoking.

NINE

THE ATTACK ON THE ELDERLY:
NURSING HOME NAZIS

They are swept away and forgotten. But the plight of the 3.5 million people who live in America's seventeen thousand nursing homes is enough to break your heart. Their stories sear through our vision of America as a kind and just society—especially for the one family in six who have a relative in a home.

Patients in many of these government-funded facilities are routinely abused, neglected, starved, dehydrated, verbally assaulted, physically harassed, robbed, unnecessarily doped and restrained, raped, tortured, and beaten.

In the coming years, more than one-fifth of us will die in a nursing home bed. We would do better to die in jail. Some of these homes constitute an American gulag, run for profit by heartless and evil people.

OFF WITH THEIR HEADS!

Nine nursing homes out of ten in this country are inadequately staffed, according to a recent federal study. A majority is so short of staff that the lives of the residents are endangered.

But understaffing isn't the worst problem. More troubling still, what staff they have too often abuses the residents. In one nursing home in three, instances of abuse have been reported—slapping, punching, kicking, cursing, raping, and even torturing of the elderly. In 10 percent of all nursing homes, the U.S. House of Representatives Committee on Government Reform reported, "The abuse violations were serious enough to cause actual harm to residents or to place the residents in immediate jeopardy of death or serious injury." Abuse complaints have tripled in the past six years.

These nursing homes that abuse or neglect residents vent their cruelty on the most helpless adults among us. Yet we pay for the vast bulk of the often harsh and inadequate care they mete out. Medicaid and Medicare fund three-quarters of nursing home care, spending $58 billion each year on this frequently vicious treatment of our elderly. Residents and their families spend $34 billion more of their own funds. That comes to over $50,000 per patient, per year. One would imagine that such financial investment would ensure considerably more compassionate, attentive, and loving care.

Steve Vancore, a spokesman for Wilkes & McHugh (a law firm that has gained nationwide attention for its advocacy on behalf of nursing home residents) sums it up best: "The nursing home industry began as a welfare program for the elderly. As such, it has become a quasi-governmental entity that is a failed experiment."

He likens the widespread abuse, so often covered up behind closed doors, to the orphanages of Charles Dickens's era in Britain. The homes, usually run as for-profit businesses, "get paid regardless of the quality of care." And, Vancore notes, "They get guaranteed clients and revenues." In addition, because the government limits the number of nursing home beds, they are assured of little competition.

Each year in the United States 2.2 million people die—and five hundred thousand of us do so in nursing homes. That so many of us will meet our end amid the squalid, unsafe, abusive, unsanitary, impersonal, and dehumanizing conditions of America's nursing homes is a travesty and an outrage.

How do nursing homes get away with offering inadequate care at best—and abusive treatment at worst—to our mothers and fathers? Why aren't they drowning in a tidal wave of lawsuits?

Because, in our particular, predatory, pecuniary way of measuring the value of human life, it is cheaper to give bad care and accept the occasional lawsuit than to offer the quality care our elderly need. It makes economic sense for these nursing homes simply to settle the claims, pay the damages, and move on.

How are those damages measured, in the jaundiced eyes of our legal system?

The process is draconian. What is the quantifiable economic loss, it asks, when an eighty-five-year-old nursing home patient dies? Is there lost future income? No. Does the patient require a lifetime of care? No—she's dead. Are the years that have been stolen from her of high enough quality to merit a lot of money? No—after all, she's in a nursing home.

What about compensatory damages? The family paid no medical bills; the government was paying for her care. How much is the pain and physical suffering of an old lady, with no more than two or three more years to live, worth in the eyes of a jury?

Not a whole lot. $50,000? $75,000? Perhaps. But in any event, it's rarely enough for a nursing home administrator to lose any sleep over.

Well, what about punitive damages? What about making the nursing home pay for the way it treats a human being in the closing years of her life? Here, the politicians have jumped in to protect their good friends, the nursing home owners, by limiting noneconomic damages. In Florida, for example, the state legislature has capped punitive damages in nursing home cases at three times economic damages.

So even if a Florida jury wants to hang a nursing home owner out to dry for his vicious conduct, it can't call for more than three times the basic award in punitive damages. With no lost future earnings, and "only" a few years of life left had treatment been adequate, even with punitive damages the award might come to $150,000—$200,000— hardly worth the years of litigation it would take to reach such a verdict.

Now, the U.S. Supreme Court has begun to hint that it wants to limit punitive damages, fixing them in sliding proportion to economic damages. In its recent decision in a suit against State Farm Insurance, the court ruled that punitive damages out of proportion to economic damages should be set aside. This decision, if applied to nursing home litigation, will make the courts practically useless in enforcing humane standards of care.

In Mississippi, the politicians have gone even further, limiting punitive damages to 4 percent of a nursing home's net worth. Apparently they're anxious not to inconvenience those who assault the elderly.

For America's nursing home owners, it's a lot cheaper to continue to provide bad care, and make sure the politicians are paid off at election time, than to provide the basic, humane care they're hired to give.

OFF WITH THEIR HEADS!

Abuse at America's Nursing Homes

The following stories, unearthed from the depths of the living hell to which we consign 3.5 million of our parents, are hard to believe. Yet they have a revolting ring of truth to them. They come not from an Edgar Allan Poe novel but from the research of the House of Representatives Committee on Government Reform:

- An eighty-year-old Missouri stroke victim suffering from dementia and memory impairment was "locked in a bathroom, hit with a belt, dragged on his knees, and hit in the head with a book by nursing home employees." The nursing home workers then bribed a brain-damaged fifty-year-old patient to attack the older man. "Because of the [older] resident's impaired memory, family members did not learn of the abuse until another staff member . . . reported the incident."

- A worker in an Ohio nursing home "slammed" a patient into a chair, "closed off his nose with his hand to cut off his airway, pried back his thumb, verbally abused him, and let him fall to the floor." According to the House Committee report, the abusive staff member still works at the nursing home.

- In Indiana, a resident of a nursing home was killed by a fellow patient. The killer had a record of fifty instances of abusive behavior and had a criminal record, but the nursing home did nothing to protect vulnerable residents from him.

- In a Texas home, "a male resident was discovered by facility staff laying on top of a female resident with his pants and underwear off, attempting to pry her legs apart. . . . Facility staff were aware of this resident's sexually aggressive behavior, but failed to take measures to prevent abuse."

- In an Ohio nursing home, "a resident with dementia abused thirteen other residents over a ten-month period, including sexually assaulting a female resident, punching and slapping numerous residents in the face, and striking another resident with a coffee mug."

One state inspector asked about a female resident who appeared to have been sexually abused. "Maybe she fell on a broomstick," the facility's director of nursing replied.

Widespread incidents were reported of verbal abuse, including many instances of such constant and contemptuous behavior by staff that residents felt intimidated and were frightened to ask for care or assistance.

Part of the problem is that nursing homes that accommodate elderly residents often also admit younger people with mental or physical problems. Some of these younger patients have criminal histories, sometimes including sexual abuse. A nursing home that caters primarily to elderly residents may not be equipped to cope with such challenges.

According to the House Government Reform Committee report, 5,283 of America's seventeen thousand nursing homes have been "cited for an abuse violation that had the potential to cause harm between January 1999 and January 2001. . . . Over 2,500 of the abuse violations . . . were serious enough to cause actual harm to residents." In 1,601 nursing homes, the abuse did cause actual harm, even death, to its residents.

The abuse of nursing home residents in our time recalls the worst stories of mental hospitals through the ages. The sexual abuse rivals the most dismal, hard-time prisons, and the physical punishment is on a par with the practices of nineteenth-century juvenile reformatories. Yet all these episodes—and thousands like them—took place in American, government-funded nursing homes between 1999 and 2001.

Abuse at nursing homes is "not reported promptly," according to a March 2002 Government Accounting Office (GAO) Study of Nursing Home Resident Abuse: "Local law-enforcement officials indicated that they are seldom summoned to nursing homes to immediately investigate allegations of physical or sexual abuse. Some of these officials indicated that they often receive such reports after evidence has been compromised."

Nor do nursing home administrators, worried about retaining their funding, report abuse promptly to state agencies charged with administering nursing homes. The GAO study found that "about 50 percent of the notifications from nursing homes (to state agencies) were submitted two or more days after the nursing home learned of the alleged abuse."

The GAO noted that "these delays compromise the quality of available evidence and hinder investigations."

The delays in reporting abuse are, typically, a result of a pattern of deception, fear, and intimidation. In one facility, the GAO found that "a resident reported to a licensed practical nurse that she had been raped in the nursing home. Although the nurse recorded this information in the resident's chart, she did not notify the nursing home management. She also allegedly discouraged the resident from telling anyone else. Two months later, the resident was admitted to a hospital for unrelated reasons and told hospital officials that she had been raped. It was not until hospital officials notified police of the resident's complaint that an investigation was conducted."

The *New York Times* reported that GAO investigators "found several reasons" for the delays in reporting nursing home abuse:

- "Patients and their relatives are often reluctant to report abuse, because the patients fear retribution and the relatives fear the patients will be told to leave."
- "Nursing home managers are reluctant to report abuse because they fear that it will cause 'adverse publicity' or that state regulators will impose fines and other penalties."
- Nursing home employees "fear losing their jobs or recrimination from coworkers if they report abuse."
- "In some states and at some nursing homes, it is difficult to learn the correct telephone number for reporting abuse."

Neither federal nor state laws prevent abusive people from finding their way into nursing home jobs. According to the GAO, "There is no federal statute requiring criminal background checks of nursing home employees nor does CMS [Centers for Medicare and Medicard Services, the agency that runs Medicare and Medicaid] require them."

While CMS does stop nursing homes from employing people convicted of abusing residents, those who have committed "similar offenses, such as child abuse, are eligible to work in nursing homes." Even when states do try to find whether potential employees have had episodes of abuse in prior employment, "criminal background checks typically do not identify individuals who have committed a crime in another state."

If nursing homes wanted, they could ask the FBI to do a nationwide background check on prospective employees. This investigation would uncover offenses committed in other states by those a nursing home is about to hire. But the GAO reports that twenty-seven states never request background checks from the FBI, and that most other states do so only rarely.

Very seldom does abuse in a nursing home lead to a successful criminal prosecution. Cases are rarely referred to police, and when they are it is often only after evidence has disappeared. Because of the nature of the abuse, and the impaired memory of many of the victims, prosecutions cannot always find good witnesses to testify in court. The GAO said, "Our work also indicated that resident testimony could be limited by mental impairments or an inability to communicate. We noted several instances in which residents sustained unexplained black eyes, lacerations, and fractures. However, despite the existence of serious injuries, investigators could neither rule out accidental injuries nor identify a perpetrator."

With no criminal prosecutions, and state legislatures shielding nursing homes from civil lawsuits by capping damages, there is no way to hold those who abuse the elderly accountable, either criminally or civilly.

But where a state prosecutor actively focuses on protecting nursing home residents, he can do a great deal to deter and punish abuse. According to the *New York Times,* "since January, 1999, the Medicaid fraud control unit in the office of the [New York] State attorney general, Eliot L. Spitzer, has filed charges of patient abuse against 86 people. Kevin R. Ryan, a spokesman for the unit, said that 60 people had been convicted and nine had been acquitted, while two cases were dismissed and the others are pending."

Most states do little to punish nursing homes that abuse residents. While they have the power to cite the nursing homes, the GAO reported that "these deficiencies rarely result in the imposition of sanctions, such as civil monetary penalties, by state . . . agencies."

Federal investigators studied 158 cases of reported abuse in Georgia, Illinois, and Pennsylvania. In only one instance—a facility in Illinois—was the nursing home fined, and that penalty was reduced on appeal.

The result, predictably, is that nursing facilities are often home to serial abusers. Between 1999 and 2001, the House Government Reform Committee found that 1,327 nursing homes were cited for more than one abuse violation, and 305 were cited for three or more—192 had five or more abuse violations!

And even when abusive staff members are fired, most can usually find jobs at other nursing homes. The 158 cases the GAO studied involved 105 abusive nurse aides, but only twenty-one of them had their conduct noted on their records, barring them from other nursing home employment. The others were free to work at other homes with no black mark against their names.

Finally, even though nurse aides are policed by state agencies—however inadequately—other nursing home employees such as laundry aides, security guards, and maintenance workers are not. While state agencies can sometimes stop abusive nurse aides from getting new nursing home jobs in their state, they can do nothing to stop these other workers from landing new jobs, even if they have been abusive to residents. The absence of viable criminal prosecutions leaves them free to do damage time and again.

As Vancore says, "If this record of abuse had taken place at child day care centers, you can imagine the political stink that would have been raised. But because it happened to our elderly tucked away in nursing homes, it is barely noticed."

Inadequate Staffing

Even when nursing homes are not abusive, they are typically understaffed. Most homes are operating at well below the minimum staff level at which good care is possible.

In 1990, Congress ordered the U. S. Department of Health and Human Services (HHS) to study nursing home staffing. Due to red tape and administrative and funding delays, the report was not finished until August 2000, when it was issued in the closing days of the Clinton administration.

The results were shocking.

After performing a complex series of analyses to gauge how different staffing levels at nursing homes correlated with bad outcomes in resident care, the report found that the "optimum" staffing level at a nursing home was 2.9 hours of nurse aide time for each day each resident spent in the facility. Only at this level, the investigators found, were staff able to handle the basic services they are expected to provide: assistance in eating, dressing, and using the bathroom, providing regular exercise, and helping to turn over or reposition elderly residents in bed who are immobile.

With less than this level of staffing, the HHS report found that nursing home residents were more likely to "experience bedsores, malnutrition, weight loss, dehydration, pneumonia, and serious blood-borne infections."

But the HHS report found that 92 percent of all nursing homes fell below the recommended level of staffing—2.9 hours per day. Indeed, 54 percent delivered less than 2.0 hours per resident-day! The report documented a direct correlation between inadequate staff and poor care, noting that "nursing homes with low staffing levels tended to have large numbers of residents with nutrition problems and bedsores."

The report calculated that nursing homes would have to hire between 77,000 and 137,000 new registered nurses, 22,000 to 27,000 new licensed practical nurses, and 181,000 to 310,000 more nurse aides to provide the needed levels of care. It would take $7.6 billion extra per year, an 8 percent increase in nursing home payments, to do it.

The report recommended minimum national staffing standards, but President Bush said no. Instead of enacting new regulations to assure adequate staffing, the *New York Times* reported, "The Bush Administration said that it wants to publish data on the number of workers at each nursing home, in the hope that 'nurse staffing levels may simply increase due to the market demand created by an informed public.' "

Sometimes the free market works. When the government published airline on-time records, service immediately improved. Circulation of data on hospital death rates and other indices of good health service have made more health-care providers toe the mark in order to attract patients. But to claim that the same approach would work for nursing homes is fatuous.

Three-quarters of nursing home care is publicly funded. Nursing homes are simply not going to provide better care unless they get more money. With 92 percent of the homes out of compliance with optimum staff standards, what are the rest of the elderly to do? Crowd into the remaining 8 percent of the facilities?

As Donna R. Lenhoff, executive director of the National Citizens'

Coalition for Nursing Home Reform, put it: "The government admits that increasing staff to the levels recommended in the [HHS] report would improve quality, but then asserts that no action can be taken. . . . That's a very weak response."

The lack of good nursing home care is evident as inadequate staffing persists. In 1999, the inspector general of HHS reported that nursing homes were getting worse in their supervision to prevent accidents, care for pressure sores (bedsores), and proper care for activities of daily living.

The most frequent care deficiencies, according to the HHS inspector general, were in the areas of:

1. *Treatment to Prevent or Treat Pressure Sores (Bedsores)*

Almost anyone who is confined to bed will develop sores where the body presses against the mattress. The most common areas are on the backs of one's heels and the buttocks. The key to averting bedsores is to turn the patient over frequently and, where he or she is suffering from dementia, to monitor him to see that he stays in the new position. But a study by the Department of Social and Behavioral Sciences of the University of California at San Francisco indicated that two-thirds of all nursing home residents suffer from pressure sores but only 7 percent are getting special skin care. Seventeen percent of nursing homes were found to be deficient in providing adequate care to prevent bedsores.

2. *Facility Free of Accident Hazards*

Almost one nursing home in four fails to take adequate measures to prevent accidents and remove hazards from the facility. With some elderly residents suffering from impaired vision and balance, it is obviously important to minimize accident-causing conditions. But 22 percent of the nursing homes fail to do so, and statistics indicate that the situation has worsened during each of the past six years.

3. *Resident Care that Maintains/Enhances Dignity*

It is very hard for any patient to preserve a sense of his or her own humanity in a nursing home, but caregivers have recognized that helping a resident remain well groomed and properly dressed is crucial to preserving their personal dignity. It's also important to encourage residents to eat independently when they are able, and to make sure they have private space and a place to keep personal property. Perhaps most important is to treat the elderly people who live in nursing homes with respect in speaking to them and listening to what they have to say. With understaffing, this ideal of care is often unattainable. Sadly, studies show that 17 percent of nursing homes were repeatedly deficient in this area in 2001, and services that promote patient dignity have seen steady erosion in the past five years.

4. *Housekeeping and Maintenance*

Seventeen percent of all nursing homes fail to maintain a sanitary, orderly, and comfortable living environment. Forcing people to live amid filth can only degrade them and undermine their sense of self-worth. Trend lines show a steady reduction over the past six years in housekeeping and maintenance.

Almost one-third of all nursing homes fail to maintain sanitary conditions in the storing, preparation, distribution, and service of food. The elderly do not recover easily from food-borne illness, yet the proportion of nursing homes found substandard in this area has risen from 22 percent in 1997 to 32 percent in 2001.

As a *U.S. News & World Report* study found, "It's not uncommon for [nursing] homes to spend only $2 or $3 per day to feed patients. At the Evergreen Gridley Health Care Center in North California, for example, the home spends an average of $1.91 daily per resident."

5. *Observing Residents' Rights to Freedom of Movement*

Too often, nursing home residents are unnecessarily restrained, denying them full or normal freedom of movement. Studies of nursing homes have shown that 11 percent regularly impose physical restraints for discipline, or even for staff convenience, when they are not medically necessary. The aides in an understaffed nursing home may find it easier to tie a patient up than to monitor him to assure that he doesn't get out of bed and wander in the halls.

Nursing homes must provide ongoing activities to meet the physical, mental, and social needs of their residents. To fight senility and dementia and to keep spirits and morale as high as possible, homes should provide ways to accommodate the interests of their residents. Homes do better in this area than in many categories. In 2001, only 8 percent fell short according to monitoring studies.

6. *Freedom From Unnecessary Drugs*

It must be very tempting to some to keep people in nursing homes over-medicated and unnecessarily sedated to make life easier for overworked nurse aides in understaffed facilities. To help residents keep their dignity and will to live, nursing homes must work to reduce unnecessary drug use. Studies show that one in ten nursing home residents is over-medicated.

7. *Appropriate Treatment for Incontinence*

Nothing saps a person's sense of dignity more than incontinence. It is critical that nursing homes provide services for people whose bladder functioning is deficient—yet studies show that one nursing home in ten fail to provide adequate services to those who need them to cope with incontinence.

8. *Activities of Daily Living Care Must Be Provided for Dependent Residents*

Often nursing home residents cannot perform the basic daily routines that we all need for proper living. Specifically, they frequently need help with nutrition, grooming, and personal and oral hygiene. The need for assistance with these basic activities is, after all, the most basic reason to enter a nursing home in the first place. But 13 percent of nursing homes fail to provide adequate care for daily living.

Given these deficiencies in patient care, it should come as no surprise that a recent study by the *St. Louis Post-Dispatch* reported that more than four thousand nursing home residents died from bedsores, malnutrition, or starvation in 1999. (And the newspaper took at face value the official reason listed on death certificates; the real number is undoubtedly higher.)

Even when death certificates list other causes of death, closer investigation finds that neglect may have been the real cause. In Little Rock, Arkansas, Coroner Mark Malcolm examined one hundred questionable nursing home deaths between 1993 and 1999. Why had they died? "I didn't know. Nobody knew," said Malcolm. "These poor souls were cremated or buried with no one but the nursing home or its doctor deciding why they died—whether it was natural causes or because they weren't properly cared for."

According to the *St. Louis Post-Dispatch,* Malcolm "interviewed families and nurses and examined whatever clinical information was available on the deceased, including medical records and nursing home charts. Seven bodies were exhumed. Working with the medical examiner's office, Malcolm determined that more than 30 percent of the death certificates listed an incorrect cause of death."

Malcolm said: "The families were being told by the nursing homes that their loved ones died of heart attacks, strokes, and other natural causes, but what we actually found was that about a third were wrongful and preventable deaths, either caused by or exacerbated by dehydration, malnutrition, including choking, or from sepsis from bedsores."

The experience of seventy-seven-year-old Indiana woman is, unfortunately, typical. A diabetic who had lost a foot, she needed periodic kidney dialysis treatments. During one such session, she became disoriented and had to sit down. Unsupervised at the hospital and unrestrained in the chair, she got up and wandered around the hallways. Predictably, she fell and broke her hip. After the hip was set, she went to a nursing home to recuperate. Exhibiting signs of dementia, she did not reposition herself in bed and had to be turned over every three hours to prevent bedsores. Because of understaffing, the nurse aides frequently forgot to turn her or, when they did, failed to follow up to check whether she had gone back to the initial position. Sores developed, got worse, became infected—and, eventually, killed her.

With government and private sources paying $92 billion each year for patient care, where is the money going? A *U.S. News & World Report* study found that one-fifth of nursing homes spend more than 20 percent of their revenue on administrative costs. With two-thirds of all nursing homes privately owned, profit margins also eat into the quality of care.

Some states provide worse nursing home care than others. Here is a list, by state, of the number of nurse aide hours per resident-day in the nursing home. Remember, as you read this dismal chart, that the recommended standard is 2.9 hours (no state met that) and the minimum is 2.0 hours (Iowa, Tennessee, Nevada, Indiana, and Illinois flunked even that low standard).

RANKING OF STATES ON ADEQUACY OF NURSE AIDE STAFFING

State	#Nurse Aide Hours/Resident-Day	State	#Nurse Aide Hours/Resident-Day
Alaska	2.7	Wyoming	2.2
Maine	2.7	Arizona	2.1
Hawaii	2.6	Arkansas	2.1
North Dakota	2.6	Connecticut	2.1
Alabama	2.5	Florida	2.1
Idaho	2.5	Kansas	2.1
Delaware	2.4	Mississippi	2.1
Kentucky	2.4	New Jersey	2.1
Montana	2.4	New York	2.1
Oregon	2.4	Pennsylvania	2.1
Washington	2.4	Rhode Island	2.1
California	2.3	South Dakota	2.1
Maryland	2.3	Utah	2.1
Michigan	2.3	Virginia	2.1
South Carolina	2.3	Colorado	2.0
Massachusetts	2.2	Georgia	2.0
Missouri	2.2	Louisiana	2.0
Nebraska	2.2	Minnesota	2.0
New Hampshire	2.2	Texas	2.0
New Mexico	2.2	West Virginia	2.0
North Carolina	2.2	Iowa	1.9
Ohio	2.2	Nevada	1.9
Oklahoma	2.2	Tennessee	1.9
Vermont	2.2	Indiana	1.8
Wisconsin	2.2	Illinois	1.7

What's the lesson here? If you have to go into a nursing home, do it in Alaska!

Employment as a nurse aide or orderly is, of course, no easy job. The pay is terrible. Nurse aides typically earn $8 per hour for "exhausting, unpleasant, and often dangerous work." In 2000, *U.S. News* found that "78% of Texas nursing homes paid average hourly rates at or below $8.20—the federal poverty line for a family of four."

According to the American Health Care Association, the nursing industry trade group, turnover for nurse aides runs between 49 and 143 percent per year, depending on region, while for registered nurses it is 28 to 59 percent annually.

These workers, who care for our parents, deserve more pay and better working conditions.

For-Profit Homes: The Worst of the Lot

The honey of guaranteed government payment for nursing care, and the promise of nearly full occupancy (85 percent on average) has lured profit-hungry private investors into the nursing home business.

Two-thirds of America's nursing homes are for-profit facilities owned by private investors. But despite the conventional wisdom—that the private sector provides better care than non-profit or public facilities—when it comes to the nursing home business, private-sector care is disastrous.

Nursing homes that are run by private, for-profit companies provide demonstrably worse care than those that are not, according to a detailed study published in the September 2001 *American Journal of Public Health*.

The study found that privately owned, for-profit nursing homes were 47 percent more likely to offer substandard care than nonprofit facilities, and 43 percent more likely to provide deficient care than public nursing homes. Their staffing was 32 percent lower than in nonprofit homes, and 23 percent lower than in government-run facilities.

Two-thirds of America's for-profit nursing homes are owned by chains, largely concentrated in the Midwest and the South. The *American Journal of Public Health* study concluded: "Our results suggest that investor-owned nursing homes deliver lower quality care than do nonprofit or public facilities. Moreover, investor-owned facilities usually are part of a chain and chain ownership per se is associated with a further decrement in quality."

Why? Because for-profit homes need to make a profit, and profits cut into resident care. Most patients are covered by Medicaid or Medicare—and reimbursement rates are the same, no matter who owns the home. The result: less money for staff in homes where profit comes first.

America's top nursing home chain, for example, made a profit of $5.28 per resident-day in 1997. The study noted that this would be enough, "at prevailing wages, to close half of the staffing gap between for-profit and nonprofit nursing homes."

Evidence of the defects in for-profit homes continues to pile up. They have higher death rates and postoperative complication rates than non-profit homes. They get lower-quality scores, and spend less on their patients while more of their expenditures go toward administrative costs.

Sometimes nursing home owners indulge in self-dealing practices that invite the suspicion that they're inflating their expenses for self-serving reasons. *U.S. News & World Report* noted that the owner of a 314-bed facility in Hollis, New York, paid herself $1.2 million in consulting fees in 1999 and an additional $629,000 in 2000.

At another nursing home in Jamaica Estates, New York, the owner paid himself $2.8 million in rent and $4.9 million in management fees in 2000, according to *U.S. News*. These owners had to pay New York State more than $11 million in back Medicaid payments "improperly made to them" over the previous five years.

The *U.S. News* study found that "self-dealing is widespread" among nursing home owners. "About seven out of every ten homes engage in these kinds of transactions."

The U.S. Department of Justice has also noted that "a number of highflying nursing home chains appear to have incorporated defrauding Medicare as part of their business strategy."

A GAO audit found massive overbilling at nursing homes. One example cited was a speech therapist "whose pay was only $12–$25 an hour, but whose company billed Medicare for $600 or more per session." That may or may not be illegal, but it sure is outrageous.

Yet even as owners of these facilities pile up profits, the care of the elderly drops. *U.S. News & World Report* describes it vividly:

It's the smell that lingers. Leave the Shields Nursing Center here [in El Cerrito, California] or its sister facility in the city next door, and the odor of stale urine and feces trails along, hovering on the sleeve. The 130 residents who live in these low-slung buildings hugging the shore of San Francisco Bay are especially sick and dependent on government payments, records show. Financial reports show the homes are about break even. But while the patients endure daily indignities and even unsafe conditions, the home's operator, William Shields, has reaped millions to furnish an affluent lifestyle that has included a small fleet of luxury cars, a million-dollar mini-mansion, and trips to Hawaii and Lake Tahoe.

The trail of abuse by the for-profit nursing home industry tells a sad tale in man's inhumanity to man. The figures are numbing, except when one considers the human toll behind each report of abuse or larceny.

April 1999: Vencor Inc., now doing business as Kindred Healthcare, was forced to repay $90 million to the federal government for defrauding Medicare; some of its executives went to jail for their misdeeds.

May 1999: Integrated Health Service CEO Robert N. Elkins was listed by *Forbes* as the most overpaid executive in America, earning $44 million over the previous five years while driving the company into bankruptcy.

February 2000: Beverly Enterprises, the nation's largest nursing home chain, pled guilty and agreed to pay a settlement of $175 million after the government claimed that they defrauded more than $460 million.

January 2001: Integrated Health Services gave Elkins a $55 million severance package, in addition to his five-year salary of $44 million, bringing his personal compensation to $99 million—all for driving a company into bankruptcy.

March 2001: Vencor Inc. agreed to repay the government $130 million of a $1.3 billion claim for "intentionally defrauding the government."

October 2001: Sun Healthcare pled no contest to felony elder abuse charges, and paid over $93,000 in fines.

December 2001: National Healthcare Corporation agreed to pay $27 million to settle allegations of Medicare fraud.

August 2002: Beverly Enterprises pled no contest to felony elder abuse charges, and paid $2 million in penalties.

OFF WITH THEIR HEADS!

Even without larceny, however, owning a nursing home is a profitable business. Dr. Susanne Seagrave, an analyst for the Medicare Payment Advisory Commission (MedPAC), testified on December 12, 2002, that "Medicare [profit] margins for all freestanding SNFs [skilled nursing facilities] average about 11 percent for fiscal year 2003." She said that those affiliated with large nursing home chains made out even better. "We do find vast differences according to whether facilities are associated with one of the top ten nursing facility chains or not. With margins for facilities in one of the top ten chains averaging about 19 percent, while margins for other facilities average about 7 percent."

Skeptical of industry claims of low profitability, *U.S. News & World Report* examined "hundreds of thousands of pages of nursing home financial statements" to assess the true financial condition of nursing homes. They found that "the nursing home industry is profitable and growing, with operators spinning a far brighter tale for Wall Street than for Capitol Hill. Many nursing homes are earning exceptionally healthy profit margins, often 20 and 30 percent."

Even as they report tough financial times in their official government filings, many nursing home operators steer big chunks of their revenues to themselves from related businesses before they calculate the bottom line.

In 1997, things took a sour turn for large nursing home chains, when President Clinton, Senate Majority Leader Trent Lott, and House Speaker Newt Gingrich agreed on a budget deal to cut federal spending. Included was a reform in the way nursing homes were reimbursed that was intended to save $9 billion.

Before the change, nursing homes had been able to pass their costs along to the government. By sending the bill to Medicare or Medicaid, they got reimbursed for whatever costs they could claim. Under pressure to cut government spending, the budget deal changed all that. Now, nursing homes must identify the diagnosis for each resident and the amount of care they require. Only then are the homes reimbursed, for what the government considers a fair amount to pay considering the residents' conditions.

This new system limited payouts and curbed the largesse of the earlier pass-along system. The cut reduced Medicare payments from an average of $268 to $243 per resident per day. That reduction may not sound like much, but it made an impact: many nursing home chains, in precarious financial condition anyway, went bankrupt.

One such was Integrated Health Systems, a chain that ran fifteen hundred nursing homes in forty-seven states (about 10 percent of the nursing homes in America). Raking in money under the old system, Integrated was getting $3 billion a year in revenue. According to SunSpot.net Business, its owner, Dr. Robert N. Elkins, "became known not only for his business strategy, but also for his lavish bonuses—$3.25 million in 1997—and perks such as his corporate jet."

When government largesse dried up, the roof caved in for Integrated. "Profitable in the second quarter of 1998," the company "lost $158 million in the third quarter" as its stock dropped from $40 a share to 1 cent by the end of 1999. Paul Willging, past president of the American Health Care Association, a nursing home trade group, said that Integrated failed, in part, because it was "leveraged up the gazoo," having borrowed $3 billion to finance its acquisitions.

Five of the nation's seven largest Medicare-dependent nursing home chains filed for bankruptcy in the wake of the Medicare payment reform. Some large nursing home chains used the bankruptcies to push for higher government funding.

U.S. News & World Report describes how they have launched "a coast to coast campaign, complete with high-level lobbying, doomsday advertising, and an RV road trip to gather petition signatures, the nations' nursing home operators are feverishly warning of disaster: the demise of their industry, and the endangerment of the lives of millions if the federal government doesn't extend billions in payments."

But the record fails to bear out their claims of penury. "In studying industry finances, the GAO recently found that the level of federal payments wasn't the problem. At the root instead were bad business decisions and higher costs—including piling up lots of debt in a campaign of pricey acquisitions."

As Andrew Gitkin, a nursing home analyst with Paine Webber, put it, "The companies that have gone bankrupt are the ones that were very aggressive with their expansion plans throughout the mid 1990s. Their balance sheets got very bloated. That put them in a vulnerable position."

It couldn't happen to nicer guys.

What We Can Do About It

The Bush administration's position—that market forces will reform the nursing home industry—is clearly misguided. An industry that gets two-thirds of its revenue from the government will never be responsive to market forces. It will march to the beat of government regulators or it will stand still.

Nor will nursing homes improve their service through productivity. The very essence of nursing care requires patience and hours. How do you improve productivity in helping an elderly person go to the bathroom or helping her eat her food or get exercise?

Bush's vaunted free-market system won't work here:

- There aren't enough vacancies in nursing homes for the elderly to have a genuine choice. Only about one nursing home bed in every seven is empty. In certain geographic areas, and among specific categories of needier elderly, the vacancy rate is likely even lower. A free-market approach must be rooted in the assumption of consumer options, but with such low vacancy rates, choice is an illusion.

- Relaxing government controls on the number of nursing home beds won't solve the problem. If more nursing homes are built and occupancy rates drop, Medicare will have to pay for the cost of building and maintaining all those empty beds, by paying more for each bed that is filled.

- The elderly people who enter nursing homes aren't at the top of their game. With impaired memory, limited verbal skills, and, in many cases, outright dementia, they aren't able to be informed consumers, reporting back to relatives on the outside how their care is falling short. Even when they are able to communicate, intimidation will play a role in keeping them from honestly sharing their burdens.

- Free-market systems are not good at enhancing human services. They work best to control costs. While cutting fraud and abuse remain priorities for government oversight of nursing homes, improvement of hands-on care has to be the top priority. Private sector systems focus obsessively on the bottom line—the financial bottom line—not on the quality of care.

- Our oldest citizens are also our least resilient and adaptable. The psychic cost of asking a vulnerable elderly resident to relocate to an unfamiliar new nursing home is likely to be prohibitively high—ruining the chances of reasonable competition.

- It is very difficult to distinguish between a good and a bad nursing home from the vantage of a potential consumer. The kind of hard data consumers need to make informed decisions about nursing home care is nearly impossible to find and, despite Bush's efforts to audit nursing facilities, is unlikely to become available. Reports of patient abuse are routinely covered up and complaints can be hard to prove in a nursing home environment.

It's going to take more money to improve nursing homes and tighter federal regulation of staffing levels. President Bush needs to overcome his aversion to regulating nursing homes—to understand the urgent demands of compassion and the need to observe the biblical commandment to "honor thy father and thy mother."

The federal government should adopt the recommended level of 2.9 hours of nurse aide care per resident-day, and require all nursing homes to live up to it as a condition of receiving federal funding.

We must also pay nurse aides and other employees decently, if we're going to attract the kind of people—and give them the kind of motivation—we would want to care for our parents.

If the available financial estimates are correct (which is rare), it would cost $7.6 billion extra annually to bring nursing home staffing up to the recommended levels. But Clinton's Medicare reforms have saved more than that. Colorado Health Care Association director Arlene Miles estimates that the new payment system has reduced nursing home reimbursement by $16 billion.

Having squeezed out some of the fraud by reforming government reimbursements to eliminate the $600-an-hour speech therapist, Washington must now reinvest some of that money in funding adequate staffing at nursing home facilities.

But if the private-enterprise solution of greater competition won't work, neither will more government regulation solve the problem on its own.

Government bureaucrats—even when they're honest—are notoriously unable to police the private sector. By the time the ponderous mechanisms of government regulation have swung into action to demand better care in a nursing home, ten other cases, crying out for intervention, are likely to have arisen to crowd it out of the spotlight.

Only by making it pay to provide good care will we be able to force our nursing homes to improve. If the profit motive works against improved patient services, care will suffer. But if it offers incentives to give quality service, care will improve.

Which brings us full circle back to litigation as the best way to assure good nursing home care. It is only the lawsuit, which so directly and quickly impacts the bottom line, that makes good care profitable and bad care profitless.

We will never be able to rely on the competence and integrity of government bureaucrats to raise nursing home standards. Nor can we trust the humanistic motivations of nursing home owners and managers to do so. But fear of lawsuits, and of the whopping damage awards that can result when angry jurors hear sordid tales of nursing home abuse, will do the trick.

Like them or hate them, trial lawyers—selfish, greedy, overreaching, and aggressive—are the only agents we can trust to improve nursing homes. In their desire to make a dollar, they will scour the homes for reports of abuse or neglect, and will invest their considerable time and talents to standing up for those who are victimized (in order to win big, fat fees for themselves).

The jury system is the best way for us to make our outrage at reports of rape, malnutrition, dehydration, bedsores, physical abuse, torture, and verbal tyranny felt in the profit centers of the nursing home industry.

With over two-thirds of all nursing homes in private, for-profit, hands, only lawsuits can sufficiently impact the financial bottom line to make good quality care profitable.

To let litigation do its work in improving the quality of care, we need to free it from the shackles the politicians and their nursing home owner buddies have imposed on the process. By blunting the ability of lawsuits to make nursing homes pay for bad or abusive care, the politicians have protected the worst among us, and denied support to the most deserving.

Legislatures in states like Florida should repeal the limitations on punitive and noneconomic damages in nursing home litigation.

The elderly who are in nursing homes have no future earning capacity. They don't have many years of life left to them even under the best of circumstances. To tie the punishment a nursing home gets for starving or torturing them to a multiple of their future earning capacity, or their economic losses, is to assure that the most vicious of owners get the gentlest of slaps on the wrist.

Legislatures should make it easier, not harder, to sue nursing homes for deficiencies in patient care, and impose special damages for instances of abuse of the elderly. The damage awards that can be reaped from nursing home litigation must be large enough to ensure that our parents are protected, and to assure that trial lawyers will take their cases on a contingency fee, so that everyone with a complaint can afford access to the justice system.

It's everyone's favorite sport to complain about "trial lawyers," and nursing home owners will do so in a heartbeat if it deflects attention from their crimes. But it isn't the trial lawyers who are abusing our parents in America's nursing homes—it's the cruel and selfish owners. And it is high time that we bring them to justice.

Whenever we tuck people away in institutions, with only bureaucrats to look after their welfare, we're likely to get the worst possible outcome for those we consign. As the condition of nursing homes in our country today demonstrates, government cannot assure quality care where human avarice tends toward abuse and neglect. We must empower residents, and their advocates, if we are to ensure that our loved ones are being cared for.

Reading through the shocking stories of inhuman neglect that riddle our nursing homes, one can only wonder how widely such problems might spread throughout the health-care industry. What about our mentally ill or handicapped? What about our children in day care centers? Are their care providers more compassionate? Is the system that oversees them any more thorough?

In examining the conditions in our nursing homes, we are meeting the results of our neglect of the elderly. We do not honor our mothers and our fathers. We tuck them away to die, out of the way, in a nursing home. And in sweeping the problem under the rug, we place the greatest value not on the quality of human life—but on silence.

We need to open the doors of America's nursing homes, and let in the air of public scrutiny. We, the children of those incarcerated, must send a message that we are vitally involved in the care of our parents, and that we will not stand for abuse or neglect.

For if we fail to hold accountable those we trust with our parents' lives, how will we live with ourselves?

EPILOGUE

It all comes down to what Lord Acton said in 1887: "Power tends to corrupt and absolute power corrupts absolutely."

The examples I attack in these chapters all reflect what happens when men and women acquire sufficient power over others that abuse becomes almost inevitable.

Sometimes the power is moral or intellectual, as with the *New York Times.* Cultivated over generations of outstanding news coverage, the newspaper has developed a reputation for fairness, integrity, and unbiased news reporting.

So how easy has it been for managing editor Howell Raines and his staff to cash in on that reputation, to inject an increasingly one-sided view of events and policies? How tempting to take a newspaper that is more believed than any other, and suddenly twist its reporting, headlines, columns, and, especially, its polls to reflect the personal views of those who control its content? The impact of corrupted power is directly proportional to the helplessness of those in its thrall. Readers of the *Times* go back day after day, no matter how biased the coverage in its pages, in the belief that there is only one paper of record—and that they must read it to be informed.

Frequently, the corrupting power is the love of political office and the popularity it brings. President Bill Clinton could have confronted al Qaeda more forcefully. He would have been successful had he forced Saddam to accept inspections to curtail his armament production. He could have thrown down the gauntlet at North Korea before it got nuclear weapons. But all these actions would have compromised his grip on the presidency, particularly when the distractions of impeachment came calling. Each might possibly have led to war and casualties, which terrified him in view of his absence of military service.

Should Clinton have declared a crisis and mobilized America to battle terror before it struck our shores? Would he have succeeded? Would he have been accused of fomenting a crisis in order to protect his grip on his office, in a manner suggested by Barry Levinson's movie *Wag the Dog*? These are all risks he should have taken—in the public interest. Instead, he chose the easy path—sweep it under the thick, plush rugs in the White House, and leave terrorism for his successor to handle. Why risk the presidency?

The Hollywood apologists, by definition, are men and women with no political power. Dime-store dissidents, they fly in the face of governmental policies and conventional wisdom and point out a different way. But they have a power, as well, and it has corrupted them as fully—cultural power.

As the actors, actresses, directors, playwrights, producers, lecturers, journalists, authors, artists, singers, musicians, philosophers, and even comedians who orchestrate what we call our social culture, their power to shape our lives is enormous. The temptation to speak up when you don't know what you are talking about must be very great. So must the pressure to conform with others in one's social set by mouthing the right words and the trendiest slogans.

Setters of cultural fashion, they impersonate politicians, just as they impersonate the characters they play, or emote their way through the lyrics songwriters have given them. All with the same implied cultural authority they maintain so imperiously in their own lines of work. But their ability to master a role—their perfect pitch, as it were—abandons them when they're challenged to interpret cold, hard facts that contradict the scripts they've already memorized.

Sometimes the "power" comes from shared history. For two centuries, France has cast herself as our most dependable ally. Now, as she moves away from us in her own selfish perception of interest, we look inward and wonder why our allies are deserting us. But the French are no longer our most reliable ally; they have become our falsest "friend," impelled to leave our side by fear, greed, and antiSemitism.

The oldest of corruptions—the lure of money—has in recent years enticed lawmakers to work with Wall Street to put our life savings at risk. The campaign contributions proved irresistible to our legislators and leaders. The big money induced them to create rules that permit, even encourage, lying and deceit by the very people we're supposed to trust to keep Wall Street honest.

It was all right to fabricate predictions as long as there was a disclaimer in small print. Accountants who helped corporate plunderers figure out how to fleece the public got away free. The politicians posed behind words like "reform" as they passed laws to give Wall Street a license to steal.

Legal power, which the political parties share in our state legislatures, is one of the most corrupting influences our society has devised. The right to draw the lines of our voting districts, and thus predetermine who will speak for us in Congress, must be positively heady! Our legislators' ability to move around blocs of Democratic, Republican, and Independent voters at will to determine the fate of our elected officials must vindicate a lifetime of obsequious service to political bosses and superstars. How wonderful it must feel for the high and mighty congressmen to come begging to lowly state legislators for district lines to assure their continued incumbency! Imagine coming down from the heights of Capitol Hill to go to the political minor leagues in Sacramento, Albany, Austin, Tallahassee, and other such venues to beg for mercy! The corrupting power to control our votes and the destiny of all America's congressmen has brought out the worst in our state legislators. . . . Who'd have thought it possible?

Then there is that most evil power of all—to have so fully sold out to diabolic forces that one works in the pay of death. Tobacco company owners, and those who work for them, deal in this purest form of evil. They are our corporate bin Ladens.

Sometimes one meets a businessman who keeps making money long after he has stopped needing it. Usually the desire to stay in the game is motivated by the excitement of the chase and the joy of competition. Frequently it is a by-product of a workaholic addiction.

But when a rich man continues to want to enrich his coffers by turning out a poisonous product that captures, addicts, and kills children, may God have mercy on his soul.

For nursing home operators, the power is physical. With more than 3 million helpless men and women under their sway, they can run their private little domains without complaint, risk, or harassment. So what if their aides beat up patients? Who'll stop them?

Who cares if staffing is inadequate and patients die of bedsores, starvation, and dehydration in twenty-first-century America? Who will complain?

If the abused and tortured sue, these evil men and women trust that the court system will offer these victims no redress. If their own consciences don't bother them, the courts can't sue, and the bureaucrats won't.

OFF WITH THEIR HEADS!

ACKNOWLEDGMENTS

First I would like to thank my wife, Eileen McGann, for her ideas and creative input. About half of the subjects covered in this book were her ideas. I get the credit, but she does much of the thinking. Her suggestions led to the chapters on the *Times*, the media, France, and Hollywood—all the good stuff.

Second, thank you to my sister-in-law Mary McGann and my niece Katie Maxwell, a student at Barnard, for the research that underpins this book. I am grateful indeed for their long-suffering patience as they gave me the facts, quotes, and sources on which this work is based.

Cal Morgan is a great, great editor whose contributions range from colorful words to profound ideas. His enthusiasm for the book helped to motivate me and his suggestions were always central and good.

A specific note of thanks to Steve Vancore from the law firm of Wilkes & McHugh for his information about the plight of nursing home residents. His searing anger at injustice is contagious.

Steve Bozeman, one of the lead antitobacco attorneys, helped to point me in the right direction as I studied the misuse of antismoking money by the nation's governors.

Thanks also to Bill Kristol, one of the brightest of commentators, for putting me onto the scandal that reapportionment has become.

My new friend, Skip Weitzen, helped a lot with the chapter on France. For this courageous battler against anti-Semitism, it was a labor of love for which I am grateful.

Thank you also to Jennifer Suitor, Lina Perl, Cassie Jones, Kurt Andrews, Joyce Wong, and Tom Wengelewski at HarperCollins for helping in the publication and promotion of this book.

Above all, thanks to Judith Regan. Her instinct for what makes a good book resembles the reliability with which the needle of a compass finds true north. Hers is a rare talent. She is so attuned to the public mood that she might have made a good living in my former profession, if she wanted to take a pay cut.

NOTES

Introduction

Page

ix   "hijacked our grief": Alice Klein, NowToronto.com, "U.S. Takes on Peace, http://www.nowtoronto.com/issues/2002-10-31/news_feature.php (2/25/03).

x   "conscience does make...": William Shakespeare, *Hamlet.* [act iii, scene i, line 83]

xvi   "$9 billion a year...": State Tobacco Settlement, Campaign for Tobacco-Free Kids, http://www.tobaccofreekids.org/reports/settlements/(2/15/03).

xvi   "lung cancer deaths dropped 14 percent...": Comprehensive Statewide Tobacco Prevention Programs Save Money, Campaign for Tobacco-Free Kids, http://www.tobaccofreekids.org/research/factsheets/pdf/0168.pdf (1/21/03).

xvii "They keep 3.5 million...": Christopher H. Schmitt, USNEWS.com, "The New Math of Old Age—Why the Nursing Home Industry's Cries of Poverty Don't Add Up," Health & Medicine. 9/30/02, http://www.usnews.com/usnews/nycu/health/articles/020930/30homes.htm.

xvii "The federal government estimates...": Summary of August 2000 HCFA Report, "Appropriateness of Minimum Staffing Ratios in Nursing Homes," Executive Summary.

xvii "Moreover, the two-thirds...": Charlene Harrington, Ph.D.; Helen Carillo, M.S.; Valerie Wellin, B.A.; Baleen B. Shemirani, B.A., "Nursing Facilities, Staffing, Residents, and Facility Deficiencies, 1995 through 2001," Department of Social and Behavioral Sciences-University of California, August 2002.

Chapter 1

3   "flooding the zone . . .": quoted in Seth Mnookin "The Changing Times," *Newsweek*, December 9, 2002, p. 46.

4   "Raines is overt . . .": Ken Auletta, "The Howell Doctrine," *The New Yorker*, June 10, 2002.

4   "The *Times* has assumed . . .": quoted in Mnookin, "The Changing Times."

4   "Many people around . . .": Ibid.

5   "certainly a shift . . .": quoted in Ibid.

5   "its daily circulation of 1.1 million . . .": Circulation of the New York Press, http://www.newyorkpress.net/NYPcirc.html (2/11/03).

6   "follow the news . . .": Mickey Kaus, "Enron's Got Nothing on Rick Berke!—The *New York Times*' Hyped-up Poll Story," http://slate.msn.com/?id=2061236 (1/13/03).

7   "cut his teeth at a time . . .": quoted in Mnookin, "The Changing Times."

7   "Target selection is key . . .": Auletta, "The Howell Doctrine."

8   "Lies, Damn Lies . . .": "Current FAQS," Quote . . . .Unquote, http://www1c.btwebworld.com/quote-unquote/p0000149.htm.

8   "A close evaluation . . . Dec 7–10 . . .": CBS/*New York Times* Poll, "Three Months After the Attacks," http://www.cbsnews.com/htdocs/pdf/bckdec.pdf (10/28/02).

10   "The *New York Times* weights its polls . . .": CBSNEWS.com, "Poll: Doubts on Military Tribunal," December 11, 2001, http://www.cbsnews.com/stories/2001/12/11/opinion/main320935.shtml (10/28/02). "Jan 21–24 . . .": CBS/*New York Times* Poll, "Enron: The Fallout Continues," http://www.cbsnews.com/htdocs/pdf/enron_poll.pdf (10/28/02). CBSNEWS.com, "Poll: Enron Fallout Rising," January 26, 2002, http://www.cbsnews.com/stories/2002/01/25/opinion/main325699.shtml (10/28/02). "July 13–16 . . .": CBS/*New York Times* Poll, "The Market, the Economy and the Scandals," http://www.cbsnews.com/htdocs/c2k/poll0717_back.pdf (10/28/02). CBSNEWS.com, "Poll: Economy Worries on the Rise," July 17, 2002, http://www.cbsnews.com/stories/2002/07/17/opinion/polls/main515481.shtml (10/28/02). "September 2–5 . . .": CBS/*New York Times* Poll, "September 11th: One Year Later," http://www.cbsnews.com/htdocs/c2k/9-11_national.pdf (10/28/02). CBS NEWS.com, "Poll: America, A Changed Country," September 7, 2002, http://www.cbsnews.com/stories/2002/09/07/september11/main521173.shtml (10/28/02). "October 3–5 . . .":

CBS/*New York Times* Poll, "The November Elections," http://
www.cbsnews.com/htdocs/c2k/pol106.pdf (10/28/02). CBS NEWS.
com: "Poll: What Voters Care About," October 6, 2002, http://
www.cbsnews.com/stories/2002/10/07/opinion/polls/main524516.
shtml (10/28/02).

12   "All of us . . .": John Zogby, e-mail correspondence with author,
February 20, 2003.

14   "Themes of the *Times*'s Post-9/11 Coverage": Based on author survey and analysis of the *New York Times* for that period.

15   "the newspaper ran twenty-five front-page . . .": Ibid.

15   "Washington and Nation Plunge . . .": R.W. Apple Jr., "Awaiting
the Aftershocks—Washington and Nation Plunge into Fight with
Enemy Hard to Identify and Punish," *New York Times*, September
12, 2001, p. 1.

15   "Scarcity of Afghanistan . . .": Michael R. Gordon, Eric Schmitt
and Thom Shanker, "Scarcity of Afghanistan Targets Prompts U.S.
to Change Strategy," *New York Times*, September 19, 2001, p. 1.

15   "Peacetime Recruits Getting . . .": Dana Canady, "Peacetime
Recruits Getting Ready for War's Perils," *New York Times*, September 20, 2001, p. 1.

15   "The Tough Afghan Terrain . . .": David Rohde, "Tough Afghan
Terrain," *New York Times*, September 26, 2001, p. 1.

15   "A Guerrilla War . . .": Serge Schmemann, "PRESIDENT SAYS
U.S. IS IN 'HOT PURSUIT' OF TERROR GROUP—A 'GUER-
RILLA WAR' Allied Units Reportedly Entered Afghanistan—No
U.S. Word," *New York Times*, September 29, 2001, p. 1.

15   "Ground Raids Seen . . .": "GROUND RAIDS SEEN AS LONG
AND RISKY—British Military Chief Expects Tough Commando
Conflict," Michael R. Gordon, *New York Times*, October 25,
2001, p. 1.

15   "Insertion of Ground . . .": Michael R. Gordon, "A Direct Engage-
ment—Insertion of Ground Troops Demonstrates Willingness to
Risk American Casualties," *New York Times*, October 20, 2001,
p. 1.

15   "Allies Preparing for . . .": Michael R. Gordon, "ALLIES
PREPARING FOR A LONG FIGHT AS TALIBAN DIG IN—
Optimism of Early October Fades—Eventual Victory Is Seen,"
New York Times, October 28, 2001, p. 1.

16   "Do you think . . .": CBS/*New York Times* Poll, "Going to War?"
September 20–23, 2001, http://www.cbsnews.com/htdocs/pdf/
poll_92401.pdf (10/28/02).

18   "In the thirty-three days . . ." Author survey of the *New York Times*.

18   "It is quite possible that America . . .": Clyde Haberman, "NYC; Diallo, Terrorism and Safety vs. Liberty," *New York Times*, September 13, 2001.

19   "After the Attacks . . .": Somini Sengupta, "AFTER THE ATTACKS: RELATIONS; Arabs and Muslims Steer Through an Unsettling Scrutiny," *New York Times*, September 13, 2001.

19   "Broader Spy Powers . . .": BUSH TELLS THE MILITARY TO 'GET READY'; BROADER SPY POWERS GAINING SUPPORT," *New York Times*, September 16, 2001, p. 1.

19   "75 in Custody . . .": Philip Shenon and Robin Toner, "75 in Custody Following Terror Attack Can Be Held Indefinitely," *New York Times*, September 16, 2001, p. 1.

19   "Senate Democrat Opposes . . .": Neil A. Lewis and Philip Shenon, "A NATION CHALLENGED: SAFETY AND LIBERTY; Senate Democrat Opposes White House's Antiterrorism Plan and Proposes Alternative," *New York Times*, September 20, 2001.

19   "Americans Give in to . . .": Sam Howe Verhovek, "AMERICANS GIVE IN TO RACE PROFILING—Once Appalled by the Practice, Many Say They Now Do It," *New York Times*, September 23, 2001.

19   "In Patriotic Time . . .": Bill Carter and Felicity Barringer, "In Patriotic Time, Dissent Is Muted—Debate Grows over Balancing Security and Free Speech," *New York Times*, September 28, 2001.

19   "We do not want the terrorists . . .": Philip Shenon and Robin Toner, "75 in Custody Following Terror Attack Can Be Held Indefinitely," *New York Times*, September 16, 2001, p. 1.

19   "from November 9 to 30 . . .": Author survey of *New York Times*.

19   "Longer Visa Waits . . .": Neil A. Lewis and Christopher Marquis, "Longer Visa Waits for Arabs; Stir over U.S. Eavesdropping," *New York Times*, November 10, 2001.

19   "U.S. Has Covered . . .": Jacques Steinberg, "U.S. Has Covered 200 Campuses to Check Up on Mideast Students," *New York Times*, November 12, 2001.

20   "White House Push . . .": Robin Toner and Neil A. Lewis, "White House Push on Security Steps Bypasses Congress—New Executive Orders—Administration Urges Speed in Terror Fight, but Some See Constitutional Concern," *New York Times*, November 15, 2001.

20   "Civil Liberty vs. . . .": Robin Toner, "Civil Liberty vs. Security: Finding a Wartime Balance," *New York Times*, November 18, 2001.

20   "On November 21, the . . .": Fox Butterfield, "A Police Force Rebuffs F.B.I. on Querying Mideast Men," *New York Times*, November 21, 2001.

20   "Police Are Split . . .": Fox Butterfield, "Police Are Split on Questioning of Mideast Men—Some Chiefs Liken Plan to Racial Profiling," *New York Times*, November 22, 2001.

20   "Bush's New Rules . . .": Dan Barry, "Bush's New Rules to Fight Terror Transform the Legal Landscape," *New York Times*, November 25, 2001, p. A1.

20   "Tact amid Rights . . .": Jodi Wilgoren, "Michigan 'Invites' Men from Mideast to Be Interviewed—Tact amid Rights Debate—Justice Dept. Instructs Local Officials on the Question That Should Be Asked," *New York Times*, November 27, 2001, p. A1.

20   "Al Qaeda Link . . .": David Firestone and Christopher Drew, "Al Qaeda Link Seen in Only a Handful of 1,200 Detainees," *New York Times*, November 29, 2001, p. A1.

20   "Groups Gird for . . .": William Gaberson, "Groups Gird for Long Legal Fight on New Bush Anti-Terror Powers," *New York Times*, November 30, 2001, p. A1.

20   "Religious and Political . . .": David Johnston and Don Van Natta Jr., "Ashcroft Weighs Easing F.B.I. Limits for Surveillance—Rules Existed Since 70's—Religious and Political Groups in U.S. Could Again Be Fair Game Under New Plan," *New York Times*, December 1, 2001, p. A1.

20   "Few in Congress . . .": Robin Toner, "Few in Congress Questioning President over Civil Liberties," *New York Times*, December 5, 2001, p. A1.

20   "Man Held Since . . .": David Johnston and Philip Shenon, "Man Held Since August Is Charged with a Role in Sept. 11 Terror Plot," *New York Times*, December 12, 2001, p. A1.

21   "Cleared After Terror . . .": Tamar Lewin, "Cleared After Terror Sweep, Trying to Get His Life Back," *New York Times*, December 28, 2001, p. A1.

21   "Even the newspaper's own poll . . .": CBS/*New York Times* Poll, "Poll: Doubts on Military Tribunal," December 11, 2001, http://www.cbsnews.com/stories/2001/12/11/opinion/main320935.shtml, http://www.cbsnews.com/htdocs/pdf/bckdec.pdf (10/28/02).

21   "Thirty-one percent said they had heard . . .": CBS/*New York Times* Poll, "Poll: Doubts on Military Tribunal," December 11, 2001, http://www.cbsnews.com/htdocs/pdf/bckdec.pdf (10/28/02).

21   "Public Is Wary . . .": Robin Toner and Janet Elder, "Public Is Wary but Supportive on Rights Curbs," *New York Times,* December 12, 2001.

21   "Americans are willing . . . are wary of . . .": CBS/*New York Times* Poll, "Poll: Doubts on Military Tribunal," December 11, 2001, http://www.cbsnews.com/htdocs/pdf/bckdec.pdf (10/28/02).

23   "481 were reported . . .": "Hate Crimes: Hate Crimes Against Arabs Surge, FBI Finds," http://loper.org/~george/trends/2002/Nov/56.html (2/15/03).

23   "In the nine months . . .": Author survey of *New York Times.*

23   "Bill O'Reilly, my gutsy colleague . . .": Eric Boehlert, "The Prime-Time Smearing of Sami Al-Arian," Salon.com, January 19, 2002, http://www.salon.com/tech/feature/2002/01/19/bubba/.

24   "death to Israel . . .": Editorial, "Protecting Free Speech on Campus," *New York Times,* January 27, 2002.

24   "bluntly suggested that . . .": Boehlert, "The Prime-Time Smearing of Sami Al-Arian."

24   "After the show . . .": David Tell, "The *Times* and Sami Al-Arian," *The Weekly Standard,* March 15, 2002.

24   "academic freedom . . .": Editorial, "Protecting Free Speech on Campus," *New York Times,* January 27, 2002.

24   "a mockery of free speech . . .": Ibid.

24   "Putting Us to . . .": Nicholas D. Kristof, "Putting Us to the Test," *New York Times,* March 1, 2002.

25   "nearly one hundred times . . .": Tell, "The *Times* and Sami Al-Arian."

25   "Behind the Rage . . .": Nicholas D. Kristof, "Behind the Rage," *New York Times,* April 16, 2002.

25   "Then the FBI . . .": ABC Action News, "After Months of Wrangling, USF Fires Suspected Professor," February 26, 2003, http://www.abcactionnews.com/stories/2003/02/030226usf.shtml and CNN.com, "FBI Charges Florida Professor with Terrorist Activities," February 20, 2003, http://www.cnn.com/2003/US/South/02/20/professor.arrest/.

25   "According to the . . .": ABC Action News, "After Months of Wrangling, USF Fires Suspected Professor," February 26, 2003, http://www.abcactionnews.com/stories/2003/02/030226usf.shtml.

25   "Between January and April . . .": Author survey of *New York Times.*

25   "In January alone, the paper ran . . .": Ibid.

26   "Democrat Assails Bush . . .": Alison Mitchell, "Democrat Assails Bush on Economy—Daschle Says Tax Cut Caused a 'Dramatic Deterioration,'" *New York Times,* January 5, 2002, p. A1.

26   "Enron Contacted Two . . .": Elisabeth Bumiller, "Enron Contacted 2 Cabinet Officers Before Collapsing—No Aid Offered, They Say—Ashcroft Recuses Himself—Auditor Says It Destroyed Company Documents," *New York Times,* January 11, 2002, p. A1.

26   "Enron Sought Aid . . .": Richard W. Stevenson, "Enron Sought Aid of Treasury Dept. to Get Bank Loans—Under Secretary Did Not Intervene Despite Calls, His Aide Says," *New York Times,* January 12, 2002, p. A1.

26   "Parties Weigh Political . . .": Richard L. Berke, *New York Times,* January 12, 2002, p. A1.

26   "Poll Finds Enron's . . .": Richard L. Berke and Janet Elder, "Poll Finds Enron's Taint Clings More to G.O.P. Than Democrats—Economy, Not Terror, Is Now Perceived as Highest Priority," *New York Times,* January 27, 2002, p. A1.

26   "Americans perceive Republicans . . .": Ibid.

26   "When it comes to their dealings . . .": *New York Times*/CBS News Poll, "Enron: The Fallout Continues," http://www.cbsnews.com/htdocs/pdf/enron_poll.pdf, http://www.cbsnews.com/htdocs/pdf/war_poll.pdf (10/28/02).

26   "a majority of Americans say . . .": Berke and Elder, "Poll Finds Enron's Taint Clings More to G.O.P. Than Democrats—Economy, Not Terror, Is Now Perceived as Highest Priority."

27   "Asked if Bush cared . . .": *New York Times*/CBS News Poll, "Enron: The Fallout Continues," http://www.cbsnews.com/htdocs/pdf/enron_poll.pdf and *New York Times*/CBS News Poll, "Complete Results," http://www.nytimes.com/packages/html/politics/20021007_POLL/021007poll-results.html.

28   "blamed . . . When you think . . .": *New York Times*/CBS Poll, "Revenge and Return," http://www.cbsnews.com/stories/2001/09/15/opinion/main311417.shtml.

28   "To build Arab support . . .": David E. Sanger, "Hard Choices for
        Bush," *New York Times,* April 1, 2002.

29   "The prospect of more violence . . .": Patrick E. Tyler, "A Rising
        Toll for Bush: No Peace, More Blame," *New York Times,* April 18,
        2002.

29   "The State Department . . .": Thomas L. Friedman, "Where the
        Buck Stops," *New York Times,* June 9, 2002, p. A15.

29   "The anti-American . . . accelerate plans for . . .": David E. Sanger,
        "U.S. to Push Harder for Political Solution in Mideast," *New York
        Times,* April 4, 2002.

30   "Between January and June . . . Between March and . . .": Author
        survey of the *New York Times.*

30   "Anger in the . . .": Neil MacFarquhar, "Anger in the Streets Is
        Exerting Pressure on Arab Moderates," *New York Times,* April 3,
        2002.

30   "Arabs' Grief in . . .": James Bennet, "Arabs' Grief in Bethlehem,
        Bombers' Gloating in Gaza—Bleeding to Death," *New York
        Times,* April 4, 2002, and Joel Brinkley, "Arabs' Grief in Bethle-
        hem, Bombers' Gloating in Gaza—Hamas Spirit Soars," *New York
        Times,* April 4, 2002, p. A1.

30   "In Nablus's Casbah . . .": James Bennet, "In Nablus's Casbah,
        Israel Tightens the Noose," *New York Times,* April 8, 2002, p. A1.

30   "Attacks Turn Palestinian . . .": Serge Schmemann, "Attacks Turn
        Palestinian Plans into Bent Metal and Piles of Dust," *New York
        Times,* April 11, 2002.

30   "Jenin Refugee Camp's . . .": James Bennet, "Jenin Refugee
        Camp's Dead Can't Be Counted or Claimed," *New York Times,*
        April 13, 2002, p. A1.

31   "Refugee Camp Is . . .": James Bennet, "Refugee Camp Is a Scene
        of Vast Devastation," *New York Times,* April 14, 2002, p. A1.

31   "For Palestinian Refugees . . .": "For Palestinian Refugees, Dream
        of Return Endures," *New York Times,* April 16, 2002, p. A1.

31   "In Rubble of . . .": James Bennet and David Rohde, "In Rubble of
        a Refugee Camp, Bitter Lessons for 2 Enemies," *New York Times,*
        April 21, 2002, p. A1.

31   "On June 30 . . .": Elizabeth Rubin, "The Most Wanted Palestin-
        ian," *New York Times,* June 30, 2002, Section 6, p. 27.

32   "Gaza Mourns Bombing . . .": John Kifner, "Gaza Mourns Bombing Victims; Israel Hastens to Explain," *New York Times*, July 24, 2002, p. A6.

33   "With five front-page . . .": Author survey of the *New York Times*.

33   "on the defensive . . .": Richard W. Stevenson and Elisabeth Bumiller, "Parties Jousting over Wrongdoing by U.S. Business," *New York Times*, July 8, 2002.

33   "in for a rude shock . . .": Richard W. Stevenson, "Old Business in New Light," *New York Times*, July 9, 2002, p. A1.

34   "212,140 shares of . . . The S.E.C. investigated . . .": Stevenson and Bumiller, "Parties Jousting Over Wrongdoing by U.S. Business."

34   "President Bush received . . .": Jeff Gerth and Richard W. Stevenson, "Bush Calls for End to Loans of a Type He Once Received," *New York Times*, July 11, 2002, p. A1.

34   "To see whether . . .": CBS/*New York Times* Poll, "The Market, the Economy and the Scandals," July 13–16, http://www.cbsnews.com/htdocs/c2k/poll0717_back.pdf (10/28/02).

34   "Poll Finds Concerns . . .": Richard W. Stevenson and Janet Elder, "Poll Finds Concerns That Bush Is Overly Influenced by Business—President Stays Popular but Pessimism Rises over Economy," *New York Times*, July 18, 2002, p. A1.

34   "The survey suggests . . .": Ibid.

36   "In July and August . . . only thirteen articles . . .": Author survey of the *New York Times*.

37   "Americans increasingly doubt . . .": Adam Clymer and Janet Elder, "Poll Finds Unease on Terror Fight and Concerns About War on Iraq," *New York Times*, September 8, 2002.

37   "By 76–22 . . . a great deal . . . a fair amount . . . confidence in the . . .": CBS/*New York Times* Poll, "September 11th: One Year Later," http://www.cbsnews.com/htdocs/c2k/9-11_national.pdf (10/28/02).

38   " 'a lot' of . . .": Ibid.

38   "83 percent . . . 82 percent . . . improving the image . . .": Ibid.

38   "But that wasn't . . .": Clymer and Elder, "Poll Finds Unease on Terror Fight and Concerns About War on Iraq," and CBS/*New York Times* Poll, "September 11th: One Year Later" (10/28/02).

39   "Profound Effect on . . .": Patrick E. Tyler and Richard W. Steven-
        son, "Profound Effect on U.S. Economy Is Seen from a War
        Against Iraq," *New York Times,* July 30, 2002, p. A1.

39   "Air Power Alone Can't . . .": Eric Schmitt and James Dao, "Air
        Power Alone Can't Defeat Iraq, Rumsfeld Asserts—Cites Secret
        Mobile Labs—Secretary Sidesteps Question of Sending in U.S.
        Ground Forces to Oust Hussein," *New York Times,* July 31, 2002,
        p. A1.

39   "Iraq Said to . . .": Michael R. Gordon, "Iraq Said to Plan Tan-
        gling the U.S. in Street Fighting—Can't Win Open Battles—Hoping
        to Frighten Washington by Raising Political Specter of House-to-
        House War," *New York Times,* August 26, 2002, p. A1.

39   "Kurds, Secure in . . .": John F. Burns, "Kurds, Secure in North
        Iraq Zone, Are Wary About a U.S. Offensive," *New York Times,*
        July 8, 2002, p. A1.

39   "Administration Seeking to . . .": Richard A. Oppel and Julia Pres-
        ton, "Administration Seeking to Build Support in Congress on Iraq
        Issue—France and Britain Press for Working with U.N.," *New
        York Times,* August 30, 2002, p. A1.

39   "Bush Asks Leaders . . .": David E. Sanger, "Bush Asks Leaders in
        3 Key Nations for Iraq Support, Little Headway Apparent—but
        Aides Say Calls Found That Russia, France and China See Baghdad
        as Threat," *New York Times,* September 7, 2002, p. A1.

39   "Rift Seen at . . .": Julia Preston, "Rift Seen at U.N. over Next
        Steps to Deal with Iraq—Bush Asks Tough Action—U.S. Distrusts
        Inspections, but Other Nations Want to Give Them a Try Before
        War," *New York Times,* September 18, 2002, p. A1.

39   "U.S. Hurries; World . . .": Todd S. Purdum, "U.S. Hurries; World
        Waits—Bush Left Scrambling to Press Case on Iraq," *New York
        Times,* September 18, 2002, p. A1.

40   "Call in Congress . . .": James Dao, "Call in Congress for Full Air-
        ing of Iraq Policy," *New York Times,* July 18, 2002, p. A1.

40   "Top Republicans Break . . .": Todd S. Purdum and Patrick E.
        Tyler, "Top Republicans Break with Bush on Iraq Strategy—Cites
        Risk of War a War Plan—Current and Former Foreign Policy Fig-
        ures Urge More Diplomatic Preparation," *New York Times,*
        August 16, 2002, p. A1.

40   "President Notes Dissent . . .": Elisabeth Bumiller, "President
        Notes Dissent On Iraq; Vowing To Listen—'Healthy Debate' on
        War—In Quick Answer to Republican Concerns, He Reserves Sole

Right to Decide Course," *New York Times,* August 17, 2002, p. A1.

40   "Bush to Put . . .": Alison Mitchell and David E. Sanger, "Bush to Put Case for Action in Iraq to Key Lawmakers—Senators Not Convinced—Powell Speaks of Differences in the Administration over Dealing with Hussein," *New York Times,* September 4, 2002, p. A1.

40   "In Senate, a . . .": Carl Hulse, "In Senate, a Call for Answers and a Warning on the Future—Focus on Iraq Criticized," *New York Times,* September 10, 2002, p. A1.

40   "Democrats, Wary of . . .": Alison Mitchell, "Democrats, Wary of War in Iraq, Also Worry About Battling Bush," *New York Times,* September 14, 2002, p. A1.

40   "Officers Say U.S. . . .": Patrick E. Tyler, "Officers Say U.S. Aided Iraq in War Despite Use of Gas—Battle Planning on Iran—New Details of 1980's Program—Help Continued as Iraqis Used Chemical Agents," *New York Times,* August 18, 2002, p.A1.

41   "Poll Finds Unease . . .": Clymer and Elder, " Poll Finds Unease on Terror Fight and Concerns About War on Iraq."

41   "Asked if they . . .": CBS/*New York Times* Poll, "September 11th: One Year Later," September 2–5, 2002, http://www.cbsnews.com/htdocs/c2k/9-11_national.pdf (10/28/02).

42   "didn't kill 3,100 . . . Osama bin Laden . . .": as quoted at CBSNEWS.com, "Bill Clinton Weighs in on Iraq," September 6, 2002, http://www.cbsnews.com/stories/2002/09/06/politics/main 521036.shtml (2/17/03).

42   "increasingly isolates the . . .": Jimmy Carter, "The Troubling New Face of America," *Washington Post,* September 5, 2002, p. A1.

42   "The war against . . .": Editorial, "Terror's Calling Card in Bali," *New York Times,* October 15, 2002, p. A26.

42   " . . . to advance debate . . ." Dean E. Murphy, "Gore, Still Coy About Plans for 2004, Calls Bush's Policy a Failure on Several Fronts," *New York Times,* September 24, 2002, p. A17.

42   "If you're going . . .": Speech by Al Gore, "Iraq and the War on Terrorism," Commonwealth Club of California, September 23, 2002, http://www.commonwealthclub.org/archive/02/02-09gore-speech.html (2/17/03).

42   "an attack on . . .": CBSNEWS.com, "Gore: Bush Attacks Civil Liberties," September 26, 2002, http://www.cbsnews.com/stories/2002/09/26/politics/main523463.shtml (2/17/03).

43   "No Rush to War": CBSNEWS.com, "Poll: No Rush to War," September 24, 2002, http://www.cbsnews.com/stories/2002/09/24/opinion/polls/main523130.shtml (10/28/02).

43   "the Bush administration . . . 27 percent . . . 64 percent": CBS/*New York Times* Poll, "September 11th: One Year Later," http://www.cbsnews.com/htdocs/c2k/9-11_national.pdf (10/28/02).

43   "51–42 . . . 71–22 . . .": CBS, "Removing Saddam Hussein," http://www.cbsnews.com/htdocs/c2k/iraqback.pdf (10/28/02).

43   "68 percent approval . . . military action . . .": CBS/*New York Times* Poll, "September 11th: One Year Later."

43   "Is Congress asking . . . 44–22 . . .": CBS, "Removing Saddam Hussein."

43   "There is broad . . .": CBSNEWS.com, "Poll: No Rush to War," September 24, 2002.

44   "296–133 . . .": "To Authorize the Use of United States Armed Forces Against Iraq, Final Vote Results for Roll Call 455," http://clerkweb.house.gov/cgi-bin/vote.exe?year=2002&rollnumber=455 (2/17/03).

44   "77–23 . . .": U.S. Senate Regulations and Records, "On the Joint Resolution (H.J.Res. 114)," http://www.senate.gov/legislative/LIS/roll_call_lists/roll_call_vote_cfm.cfm?congress=107&session=2&vote=00237 (2/17/03).

44   "Public Says Bush . . .": Adam Nagourney and Janet Elder, "Public Says Bush Needs to Pay Heed to Weak Economy—Many Fear Loss of Jobs—Poll Finds Lawmakers Focusing Too Much on Iraq and Too Little on Issues at Home," *New York Times,* October 7, 2002, p. A1.

45   "Regardless of how . . .": CBS/*New York Times* Poll, "The November Elections," October 3–5 , 2002, http://www.cbsnews.com/htdocs/c2k/pol106.pdf (10/28/02).

45   "no matter . . .": Nagourney and Elder, "Public Says Bush Needs to Pay Heed to Weak Economy—Many Fear Loss of Jobs—Poll Finds Lawmakers Focusing Too Much on Iraq and Too Little on Issues at Home."

46   "Asked in October . . . By contrast . . .": CBS/*New York Times* Poll, "The November Elections," Ocrober 3–5, 2002, http://www.cbsnews.com/htdocs/c2k/poll106.pdf (10/28/02).

46   "3.2 percent while adjusting . . . The fact that 67 percent . . .": Ibid.

46   "wary of war . . . not quite four . . .": Editorial, "A Nation Wary of War," *New York Times,* October 8, 2002, p. A30.

47   "Positive Ratings for . . .": Adam Nagourney and Janet Elder, "Positive Ratings for the G.O.P, If Not Its Policy," *New York Times,* November 26, 2002.

47   "Americans hold favorable . . .": Ibid.

47   "Mr. Bush's enthusiasm . . .": Ibid.

47   "No, voters opposed . . .": *New York Times*/CBS News Poll, "The Bush Agenda and the GOP Congress, November 20–24, 2002, http://www.cbsnews.com/htdocs/c2k/bush_back1125.pdf.

48   "Incredibly, in the entire . . .": Nagourney and Elder, "Positive Ratings for the G.O.P, If Not Its Policy."

Chapter 2

55   "R.W. Apple Jr., writing in . . .": R.W. Apple Jr., "Bush Peril: Shifting Sand and Fickle Opinion," *New York Times,* March 30, 2003, p. A1.

55   In London, the *Independent* . . .": "The Iraq Conflict: Rumsfeld Shows the Strain as Experts Query His Strategy," Rupert Cornwell, *The Independent,* March 28, 2003, p. 3.

55   "rear was exposed . . . Every general who ever . . .": James Lakely, "Television, Newspapers Wrong on War in Iraq," *Washington Post,* April 13, 2003.

55   "with every passing . . .": Apple, "Bush Peril: Shifting Sand and Fickle Opinion."

55   "As Mao famously . . .": R.W. Apple Jr., "Iraqis Learn the Lessons of How U.S. Fights Wars," *New York Times,* March 27, 2003, p. B1.

55   "Iraq's best soldiers . . .": "Diminished Expectations in Iraq," *New York Times,* March 25, 2003, p. A16.

55   "the latest evidence that . . . thousands of fedayeen . . .": Ibid.

56   "bogged down in . . .": Ibid.

56   "Forget the easy . . .": Lakely, "Television, Newspapers Wrong on War in Iraq."

56   "The cover story of *Newsweek* . . .": Ibid.

56   "some air-power advocates . . .": Ibid.

56   "Rummy was grumpy . . .": "No More Saddam TV," Maureen
        Dowd, *New York Times,* March 26, 2003, p. A17.

56   "cocky theorists of . . . when Tommy Franks . . .": Ibid.

57   "Best-laid Plans . . .": Lakely, "Television, Newspapers Wrong on
        War in Iraq."

57   "The Pentagon's worst mistake . . .": H. D. S. Greenway, "Viet-
        nam's Lessons Forgotten in Iraq," *Boston Globe,* April 4, 2003,
        p. A17.

57   "At the end of the . . .": Joseph Perkins, "Why Have the War Crit-
        ics Been So Wrong?," *San Diego Union Tribune,* April 12, 2003,
        p. B7.

57   "could take a couple . . .": Ibid.

58   "the strain is even . . .": Cornwell, "The Iraq Conflict: Rumsfeld
        Shows the Strain as Experts Query His Strategy."

58   "Did Washington, seduced . . .": Ibid.

58   "a bold and proactive . . .": "Done Right, Real-Time War Cover-
        age Benefits Public," *USA Today,* March 25, 2003, p. 16A.

59   "We need to tell the . . .": Ibid.

59   "the older generation . . .": Jim Rutten, "A News-Altering Experi-
        ence?" *Los Angeles Times,* April 12, 2003, p. E1.

59   "the line between . . .": "Media, Terrorism and War: Beyond Real-
        ity TV," *San Diego Union Tribune,* March 27, 2003, p. B13.

59   "Thank goodness that . . .": Ibid.

59   "eight in ten of those . . .": Howard Kurtz, "Critics of War Tend to
        Be Critics of Coverage," *Washington Post,* April 10, 2003, p. C09.

60   "Why is this all . . .": Maureen Dowd, "Back Off, Syria and
        Iran!", *New York Times,* March 30, 2003, p. D13.

60   "the hawks want Iraq . . .": Ibid.

60   "On April 1 . . .": Bernard Weinraub, "Rumsfeld's Design for War
        Criticized on the Battlefield," *New York Times,* April 1, 2003,
        p. A1.

60   When the North . . .": "Vietnam's Lessons Forgotten in Iraq,"
        H. D. S. Greenway, *Boston Globe,* April 4, 2003, p. A17.

61   "so far the people . . .": "Diminished Expectations in Iraq," *New York Times,* March 25, 2003, p. A16.

61   "are all Saddam . . .": Ibid.

61   "What happened to the . . .": Lakely, "Television, Newspapers Wrong on War in Iraq."

61   "an arrogant blunder . . .": Ibid.

61   "Iraqis hate the U.S. . . .": Nicolas Kristof, "The Stones of Baghdad," *New York Times,* October 4, 2002, p. A27.

61   "a people that is still . . .": Thomas Edsall, "Hawks on War Against Hussein Stay the Course," *Washington Post,* March 28, 2003, p. A34.

61   "will have to wage . . .": Lakely, "Television, Newspapers Wrong on War in Iraq."

62   "the climactic battles . . .": "Diminished Expectations in Iraq," *New York Times.*

62   "intelligence sources are . . .": Lakely, "Television, Newspapers Wrong on War in Iraq."

62   "This could be . . .": Ibid.

62   "As many people had . . .": Ibid.

62   "I think the news media . . .": Ibid.

63   "It is clear that within . . .": Howard Kurtz, "Peter Arnett, Back in the Minefield," *New York Times,* March 31, 2003, p. C01.

63   "Every day gets . . .": John Daniszewski, "Every Day Gets Worse and Worse," *Los Angeles Times,* March 27, 2003, p. A1.

63   "Can you believe it? . . .": Bill O'Reilly, "The Truth Police on Patrol over the Media Coverage," *Fox News Channel,* March 27, 2003.

63   "withheld details of Saddam's . . .": Lakely, "Television, Newspapers Wrong on War in Iraq."

63   "Among the things Jordan . . .": Michael Starr and Deborah Orin, "Fury Over CNN Big's Cover-up," *New York Post,* April 12, 2003, p. 10.

64   "The process of trying to . . .": Jim Rutenberg and Bill Carter, "Television Producers Are Struggling to Keep Track of War's Progress, or the Lack of It," *New York Times,* March 26, 2003, B14.

64   "is definitely not giving . . .": Ibid.

64   "such concerns have led . . .": Ibid.

64   "When Secretary of Defense . . .": Lucian K. Truscott IV, "In This War, News Is a Weapon," *New York Times,* March 25, 2003, p. A17.

64   "Not since the halcyon . . .": Ibid.

Chapter 3

67   *"Après moi, le deluge":* Ebenezer Cobham Brewer, the 19th-century author of *The Dictionary of Phrase and Fable,* says that the saying "après moi, le deluge" is improperly, although widely, attributed to Louis XV. He credits it to Prince Metternich of Austria, who he says originated it in a reference to how the post–Congress of Vienna European system of monarchies would never outlast his death. He adds that "the Prince borrowed it from Mme. Pompadour who laughed off all the remonstrances of ministers at her extravagance by saying '*après nous le deluge.*' "

72   "ripped a five-floor . . .": Jim McGee, "2 Held in Trade Center Blast Are Indicted." *Washington Post,* March 18, 1993, p. A4.

72   "the full measure of . . .": Robert McFadden, "Explosion at the Twin Towers: The Overview; Inquiry into Explosion Widens; Trade Center Shut for Repairs," *New York Times,* February 28, 1993, p. 1.

72   "Clinton had a full . . .": John W. Mashek, "White House Feeling Tremors From Bombing," *Boston Globe,* March 2, 1993, p. 17.

72   "to keep your . . .": Ralph Blumenthal, "Crisis at the Twin Towers: The Overview; Inquiry Is Pressed on Cause of Blast at Trade Center," *New York Times,* March 2, 1993, p. 1.

72   "while security was . . .": Mashek, "White House Feeling Tremors From Bombing."

73   "The [Clinton] Administration's . . .": Bill Gertz, *Breakdown* (Washington, DC: Regnery Publishing, 2002), p. 17.

73   "In the thirty-three . . .": David Frum, *The Right Man* (New York: Random House, 2003), p. 143.

74   "looking back . . .": Judith Miller, "A Nation Challenged: The Response; Planning for Terror but Failing to Act," *New York Times,* January 30, 2002, p. 1.

74 "Indeed, the director . . .": David Horowitz, Interview with David Corn, *Hannity and Colmes*, Fox News Channel, August 12, 2002.

78 "showed up to collect . . .": N. R. Kleinfeld, "THE TWIN TOWERS: The Suspect; More Light Is Shed on a Shadowy Life," *New York Times*, March 6, 1993, p. 25.

78 "with the exception of mastermind Ramzi Ahmed Yousef . . .": Bruce Frankel, "FBI Hunts 6th Bomb Suspect," *USA Today*, April 1, 1993, p. 3A.

78 "soon emerged as an . . .": Steven Emerson, *American Jihad* (New York: Free Press, 2003), p. 46.

78 "in the company of Ahmad M. Ajaj . . .": Ibid., p. 46.

78 "Who wrote Mr. Ajaj's . . .": Richard Bernstein, "EXPLOSION AT THE TWIN TOWERS: The Missing Pieces; Convictions in World Trade Center Trial Solve Only Part of a Big, Intricate Puzzle," *New York Times*, March 5, 1994, p. 1.

79 "Salem carried hidden microphones . . .": Emerson, *American Jihad*, p. 49.

79 "simultaneous strikes . . .": Ibid., p. 49.

79 "hoped Tower One . . .": Ibid., p. 52.

79 "also participated in a plan . . ." Ibid., p. 53.

81 "has been secretly developing nuclear weapons . . .": Peter Slevin and Karen DeYoung, "N. Korea Admits Having Secret Nuclear Arms; Stunned U.S. Ponders Next Steps," *Washington Post*, October 17, 2002, p. A01.

82 "agreed to arrange construction": Thomas W. Lippman, "Perry May Be Named to Try to Salvage Pact with N. Korea," *Washington Post*, October 4, 1998, p. A27.

82 "C.I.A.'s National Intelligence Estimate . . .": Slevin and DeYoung, "N. Korea Admits Having Secret Nuclear Arms; Stunned U.S. Ponders Next Steps."

82 "diverted nuclear fuel . . .": Daniel Williams, "U.S. Bid to Build Coalition on North Korea Is Resisted; China, Japan Opposing Sanctions Approach," *Washington Post*, June 10, 1994, p. A1.

83 "is the first time . . .": Ibid.

83 "senior administration officials . . . unlikely to initiate military action . . .": Ibid.

84   "Frankly, he was . . .": R. Jeffrey Smith and Ann Devroy, "Carter's Call from N. Korea Offered Option; Administration Seized on New Chance at Diplomacy," *Washington Post,* June 26, 1994, p. A1.

84   "lead to possible . . .": Ibid.

85   "Gore urged everyone . . .": Ibid.

85   "to allow inspectors . . .": James Sterngold, "North Korea Invites Carter to Mediate," *New York Times,* September 2, 1994, p. 7.

85   "cannot be reprocessed . . .": Smith and Devroy, "Carter's Call From N. Korea Offered Option; Administration Seized On New Chance at Diplomacy."

86   "internationally monitored containment . . .": "Accord with North Korea," *Washington Post,* October 19, 1994, p. A22.

86   "gigantic political breakthrough . . .": Ibid.

86   "North Korea's record . . . freeze and dismantle . . .": Ibid.

86   "Republicans criticized the . . .": Thomas W. Lippman, "Perry May Be Named to Try to Salvage Pact with N. Korea."

86   "subsequently insisted that . . .": Nicholas Eberstadt, "The Dangerous Korea; The Current North Korean Regime Guarantees Crisis After Crisis," *National Review,* December 31, 1998, vol. L, no. 25.

86   "it turns out they . . .": CNN transcript of *Larry King Live* with Bill Clinton, February 6, 2003.

87   "a huge secret underground complex . . .": Sanger, David, "North Korea Site an A-Bomb Plant, U.S. Agencies Say," *New York Times,* August 17, 1998, p. A1.

87   "U.S. intelligence analysts . . .": Dana Priest, "Activity Suggests N. Koreans Building Secret Nuclear Site," *Washington Post,* August 18, 1998, p. A01.

87   "Later, the North Koreans . . .": Eberstadt, "The Dangerous Korea," *National Review.*

87   "administration officials have told . . .": Lippman, "Perry May Be Named to Try to Salvage Pact with N. Korea."

88   "to condition funding on a . . .": Ibid.

88   "American negotiators who hammered . . .": Eberstadt, "The Dangerous Korea."

88   "gave it the opportunity to . . .": Larry Niksch, "North Korea's Nuclear Weapons Program," Congressional Research Service, January 22, 2003, p. 9.

89   "The INS had no organized . . .": Ashley Dunn, "Greeted at Nation's Front Door, Many Visitors Stay on Illegally," *New York Times,* January 3, 1995, p. 1.

89   "Next on his list . . .": George Stephanopoulos, *All Too Human* (New York: Back Bay Books, 2000), p. 40.

90   "Mohamed Atta . . . was stopped by police . . .": Jim Yardley, "Mohamed Atta in Close Call in Incident at Miami Airport," *New York Times,* October 17, 2001, p. B1.

90   "Nawaf Alhazmi . . . for speeding": "Okla. Trooper Cited Hijacker for Speeding Last April," *Washington Post,* January 21, 2002, p. A02.

91   "Before 9/11 there really was . . .": Jonathan Aiken, "September 11 Hijacker Stopped By Police Two Days Before Attacks," CNN transcript, January 8, 2002.

92   "Today our nation joins . . . to purge . . .": "Remarks by President, Governor and Dr. Graham at Memorial Service," *New York Times,* April 25, 1995, p. B8.

92   "The President told the victims' . . .": Sharline Chiang and Corky Siemaszko, "Church Bells Rang Out in Oklahoma City and Across the Country Yesterday as Americans Everywhere Mourned the Victims of a Terror Bombing in the Heartland," New York *Daily News,* April 24, 1995, p. 3.

92   "The anger you feel is valid . . .": "Remarks by President, Governor and Dr. Graham at Memorial Service," *New York Times.*

92   "Appearing on *60 Minutes* . . .": President Clinton discusses Oklahoma City bombing, April 23, 1995.

92   "seek new authority . . .": Todd Purdum, "Clinton Seeks Broad Powers in the Battle Against Terrorism," *New York Times,* April 24, 1995, p. 1.

93   "Under current guidelines . . .": Neil Lewis, "Clinton Plan Would Broaden F.B.I. Powers," *New York Times,* April 25, 1995.

93   "the requirement that law . . .": John Harris, "President Expands Proposal for Countering Terrorism; 1,000 New Jobs, Tagging Explosives, Military Included," *Washington Post,* p. A01.

93   "to compile information on . . .": Lewis, "Clinton Plan Would Broaden F.B.I. Powers," *New York Times.*

94   "that if the F.B.I. had gone . . . as part of an effort . . .": Gertz, *Breakdown,* p. 32.

94   "Dick wanted a 'national crusade' . . .": Stephanopoulos, *All Too Human,* p. 340.

95   "the Clinton administration began . . .": Kurt Eschewal, "Terror Money Hard to Block, Officials Find," *New York Times,* December 10, 2001, p. 1.

95   "the number of cases brought . . .": Karen DeYoung, "Past Efforts to Stop Money Flow Ineffective; Coordination of U.S. Approach May Be Key," *Washington Post,* September 25, 2001, p. A08.

95   "beginning in 1999 . . .": Eschewal, "Terror Money Hard to Block, Officials Find."

96   "these visits were not . . .": Ibid.

96   "One must question . . .": Sebastian Rotella, "Investigator Writing to the U.N. Contends That a Blurring Between Religion and Finance in the Kingdom Is Hindering Reform," *Los Angeles Times,* December 24, 2002, p. 5.

96   "a financial network called . . .": Eschewal, "Terror Money Hard to Block, Officials Find."

98   "follow terrorists as they . . .": President Bill Clinton, weekly radio address, August 10, 1996.

99   "Switzerland over the past . . .": Clinton quoted in Todd Purdum, "On Vacation, Clinton Blasts G.O.P. Critics of Terror Bill," *New York Times,* August 11, 1996, p. 23.

99   "a bizarre coalition . . .": Lally Weymouth, "Odd Alliances Against the War on Terrorism," *Washington Post,* August 14, 1996, p. A21.

100   "the same stringent . . .": Ann McFeatters, "Clinton Signs Anti-Terror Bill; Critics Rap It as Far-Reaching," Cleveland *Plain Dealer,* April 25, 1996, p. 10A.

100   "to stop Louis Farrakhan . . .": Ibid.

101   "While here, the hijackers . . .": James Risen and David Johnston, "F.B.I. Account Outlines Activities of Hijackers Before 9/11 Attack," *New York Times,* September 27, 2002, p. 1.

102   "argued that the United States . . .": David Sanger, "A Senate Bill Would Punish Iran's Foreign Oil Partners," *New York Times,* December 13, 1995, p. D5.

102   "Spurred by the decision . . .": Ibid.

102   "no basis in international . . .": Ibid.

103   "better to continue the dialogue . . .": Rick Atkinson, "Divergent Policies Toward Iran Strain U.S.-German Relations," *Washington Post*, June 27, 1996, p. A21.

103   "you cannot do business . . .": Alison Mitchell, "Clinton Signs Bill Against Investing in Iran and Libya," *New York Times*, August 6, 1996, p. 1.

104   "We are working to rally . . .": Bill Clinton, acceptance speech at the Democratic National Convention, August 29, 1996, http://www.fas.org/spp/starwars/elect96/bc960829.htm.

104   "France, Germany and . . .": Bill Nichols, "Anti-Terror Law Signed by Clinton," *USA Today*, August 6, 1996, p. 1A.

104   "French, Russian, and Malaysian . . .": Thomas Lippman, "U.S. Aides Still Divided over Sanctions on Foreign Investors in Iran," *Washington Post*, March 6, 1998, p. A33.

104   "Clinton's senior foreign policy advisers . . .": Ibid.

105   "if the United States . . .": Ibid.

105   "dual doormat . . .": William Safire, "Dual Doormat Policy," *New York Times*, January 12, 1998, p. A21.

106   "the operating theory . . .": David Nyhan, "The Road to Terror," *Boston Globe*, July 19, 1996, p. A17.

106   "The explosion appears . . .": as quoted in New York *Daily News*, June 26, 1996, p. 3.

107   "U.S. officials . . .": R. Jeffrey Smith, "Blast 'Hints' May Have Been Ignored; U.S. Officials Cite Suspicious Incidents Before Saudi Terrorist Attack," *Washington Post*, June 28, 1996, p. A1.

107   "They didn't let . . . fearful of what we might find . . .": Ibid.

108   "seek U.S. access . . .": R. Jeffrey Smith, "U.S. Requests Access to Saudi Bomb Suspects; FBI Director to Seek Assurances from King," *Washington Post*, July 13, 1996, p. A01.

108   "We cannot accept . . .": quoted in Ibid.

108   "the Saudi Arabian . . .": quoted in R. Jeffrey Smith, "FBI Still Lacks Full Saudi Cooperation in Bomb Probe," *Washington Post*, February 13, 1996, p. A26.

108 "We have not gotten . . .": quoted in Roberto Suro, "Freeh Criticizes Saudis on Bomb Probe; Kingdom Has Withheld Important Evidence, FBI Director Says," *Washington Post,* January 23, 1997, p. A08.

108 "A glimmer of what . . .": "Wealthy Saudi May Have Had Role in Khobar Bombing; an Investigation Is Under Way," CNN, 1997.

109 "one of the great burdens . . .": quoted in Brian McGrory, "G-7 Summit Leaders Vow Terror Fight; U.S. Agenda Dominates Talks After Saudi Attack," *Boston Globe,* June 28, 1996, p. 1.

109 "by successfully pushing . . .": Ibid.

109 "We will spare no . . .": President Clinton's remarks regarding the explosion at Centennial Olympic Park, July 27, 1996.

111 "We will improve airport . . .": Bill Clinton, acceptance speech at Democratic National Convention, August 29, 1996.

111 "Clinton proposed to spend . . .": Robert Davis, "$1B Anti-Terror Plan Detailed," *USA Today,* September 10, 1996, p. A1.

111 "The plan includes . . .": Ibid.

111 "not only will . . .": quoted in Ibid.

112 "based on information . . .": Don Phillips, "Aviation Safety Panel Readies Suggestions; Passenger Profiles Are Recommended," *Washington Post,* February 5, 1997, p. A11.

112 "The commission was . . .": Richard Sisk, "Profiling Ridiculed Doubts on Air Terror Plan," New York *Daily News,* September 15, 1996, p. 4.

112 "Rounding up the usual . . .": quoted in Ibid.

112 "Ultimately, the airlines voluntarily . . .": Michael A. Fletcher, "Airport Precautions Causing Insecurity; Some Minority Passengers Suffer Delays, Humiliation," *Washington Post,* February 21, 1997, p. A01.

112 "Vice President Gore's . . .": "Sensible Air Safety Goals, but When Will They Arrive?" *USA Today,* February 13, 1997, p, 10A.

112 "all the tools of . . .": quoted in Ibid.

112 "the commission instructed . . .": Ibid.

113 "the ATA used extensive . . .": Mark Green, *Selling Out: How Big Corporate Money Buys Elections, Rams Through Legislation, and*

Betrays Our Democracy (New York, NY: ReganBooks, 2002), p. 164.

115   "firm and commensurate . . .": Remarks by President Clinton from the White House, June 26, 1999.

115   "Saddam began by persuading Turkey . . .": Cayle Murphy, "Turkish-Iraqi Oil Plan Raises Sanctions Issue," *Washington Post,* April 29, 1994, p. A46.

115   "does have some elements . . .": Ibid.

116   "manipulate the oil . . .": Paul Lewis, "U.S. and Britain Object to Iraq Oil Plan," *New York Times,* April 24, 1996, p. A8.

116   "Iraq must accept . . .": Josh Goshko, "Iraq Accepts Strict Terms for Sale of Oil," *Washington Post,* May 21, 1996, p. A08.

117   "Illegally sold oil . . .": Frank Murkowski, "Our Toothless Policy on Iraq," *Washington Post,* January 25, 1999, p. A21.

117   "a source of revenue . . .": quoted in Ibid.

117   "adequate safeguards . . .": Ibid.

118   "officials monitoring the . . .": Ibid.

119   "2 aircraft carriers . . .": Francis X. Clines, "Clinton Says U.S. Will Wait and See as Iraqis Back Off," *New York Times,* November 21, 1991, p. A1.

119   "Saddam Hussein must comply . . .": quoted in Ibid.

119   "was immediately followed . . .": Ibid.

119   "There is absolutely . . .": quoted in Ibid.

119   "the impression that 'there . . .": Ibid.

119   "might not be opposed . . .": Ibid.

119   "Jim Hoagland pieced together . . .": Jim Hoagland, "Crisis-Managing in a Fog," *Washington Post,* November 26, 1997, p. A19.

120   "to allow him to . . .": Ibid.

120   "stores of deadly . . .": Clines, "Clinton Says U.S. Will Wait and See as Iraqis Back Off."

120   "should avoid sensitive sites . . .": quoted in "Iraq Insists That U.N. Arms Inspectors Avoid Sensitive Sites," Reuters/*New York Times,* November 24, 1997, p. A8.

121   "immediately and unconditionally...": Barbara Crossette, "Divided U.N. Council Approves New Iraq Arms Inspection Plan," *New York Times*, December 18, 1998, p. A1.

121   "Blair labeled the new Iraq policy...": "The 70-Hour War: War of Words," *The Independent*, December 18, 1999.

121   "You were up to the level...": quoted in Howard Schneider, "As Key Sites Lie in Ruins, a Durable Saddam Declares 'Victory,' " *New York Times*, December 21, 1998, p. A1.

122   "The distinctions between the...": Murkowski, "Our Toothless Policy on Iraq."

122   "has a very long history...": President Clinton's remarks at Georgetown University, November 7, 2001.

123   "I can imagine Bill Clinton's...": Author's conversation with Bill Curry, August 1995.

Chapter 4

128   "If there is a war...": Truthout.org, "Text of Statement by Sean Penn at News Conference in Baghdad," http://www.truthout.org/docs_02/12.17E.penn.iraq.htm (12/15/02).

128   "This war is about...": quoted in Ananova.com, "Hoffman Blasts Bush War Plans," http://www.ananova.com/news/story/sm_747771.html (2/6/03).

128   "The war mongers...": Woody Harrelson, "I'm an American Tired of American Lies," *The Guardian*, October 17, 2002, http://www.guardian.co.uk/g2/story/0,3604,813189,00.html (2/6/03).

128   "There can be no more deaths...": "A Statement of Conscience: Not in Our Name," http://www.notinourname.net/statement.conscience.html (2/24/03).

129   "an adolescent bully...": quoted in Nina Shapiro, "War or Peace? Local Notables Take Their Stand," *Seattle Weekly Online*, March 5–11, 2003.

129   "When I see an American flag...": quoted in NewsMax.com, "Director Altman: 'Americans Are Full of It!'," January 22, 2002.

130   "I've listed...about...": quoted at CommonDreams.org, Stephanie Holmes, "Author Gore Vidal Slams U.S. for Waging 'Perpetual War,' " November 24, 2001, http://www.commondreams.org/headlines01/1124-04.htm (2/24/03).

131   "cowardly attack . . .": Susan Sontag in Hendrik Hertzberg et al, "Tuesday and After," *The New Yorker,* September 24, 2001, vol. 77, issue 29, p. 27.

131   "We're puzzled over . . .": Andy Rooney, "Patriotism: Too Much of a Good Thing," *60 Minutes,* CBSNEWS.com, February 18, 2002, http://www.cbsnews.com/stories/2002/02/15/60minutes/rooney/main329517.shtml (2/6/03).

131   "Everything wrong with . . .": "IDIOCY WATCH," *New Republic,* November 26, 2001, vol. 225, issue 22, p. 8.

131   "I really believe . . .": Laurie Goodstein, "Falwell's Finger Pointing Inappropriate, Bush Says," *New York Times,* September 15, 2001.

132   "the fundamental rights": Feminist Majority Foundation, "Declaration of the Essential Rights of Afghan Women—Statement of Support," June 28, 2000, http://www.feminist.org/afghan/declarationsupport.asp.

133   "Blood on *our* hands . . .": Truthout.org, "Text of Statement by Sean Penn at News Conference in Baghdad."

133   "diverted money from . . .": "Saddam Hussein's Iraq," Prepared by the U.S. Department of State, released September 13, 1999, http://www.fas.org/news/iraq/2000/02/iraq99.htm.

134   "he found it . . .": BBC News, "Sean Penn Urges Peace with Iraq," December 16, 2002, http://news.bbc.co.uk/1/hi/world/middle_east/2577981.stm.

134   "The *New York Times* . . .": John F. Burns, "Actor Follows His Own Script on Iraq and War," *New York Times,* December 16, 2002.

134   "keep their political opinions . . .": Rick Lyman, "Celebrities Become Pundits at Their Own Risk," *New York Times,* March 2, 2003.

134   "In the first Crusade . . .": Bill Clinton, Speech at Georgetown University, November 7, 2001, http://www.georgetown.edu/admin/publicaffairs/protocol_events/events/clinton_glf110701.htm (2/7/03).

135   "until every drop . . .": Abraham Lincoln, Second Inaugural Address, March 4, 1865.

135   "This country once looked the other way . . .": Clinton, Speech at Georgetown University, November 7, 2001.

135   "4,697 fellow Americans . . .": James H. Madison, *A Lynching in the Heartland—Race and Memory in America* (New York: Palgrave/Macmillan, 2003), p. 13.

136   "George W Bush . . .": quoted in BBC News, "Sheen Slates 'Bad Comic' Bush," BBCNews Online, February 13, 2001, http://news. bbc.co.uk/2/low/entertainment/1166696.stm.

136   "Mr. Sheen, it . . .": "Who Mentored Martin Sheen?", from Matilda Cuomo, *The Person Who Changed My Life* (New York: Carol Publishing Group, 1999).

137   "Alcoholics Anonymous and jazz . . .": quoted in BBC News, "Sheen Slates 'Bad Comic' Bush."

137   "a president who . . .": CNN.com, "Mandela: U.S. Wants Holocaust," January 30, 2003, http://www.cnn.com/2003/WORLD/ meast/01/30/sprj.irq.mandela/index.html.

137   "The moral climate . . .": quoted on NewsMax.com, "CBS TV Star Compares America to Nazi Germany," February 3, 2003, http://newsmax.com/showinsidecover.shtml?a=2003/2/3/172603 (2/24/03).

138   "opposing the war . . ." "Backlash Against Celebrities," *New York Times,* March 23, 2003.

138   "a fictitious president": "Filmmaker Stirs Emotions With Anti-Bush Remarks," Tom Maurstad, *Dallas Morning News,* March 25, 2003.

138   "a simplistic and inflammatory view . . .": Burns, "Actor Follows His Own Script on Iraq and War."

139   "possessed of evil": quoted on Excite.com, "Harry Belafonte Says U.S. Leader 'Possessed of Evil,'" March 4, 2003.

139   "The government itself . . .": quoted in New York *Daily News,* "Clooney Isn't Joining Dubya's Gang," January 20, 2003, http:// www.nydailynews.com/news/gossip/story/52957p-49641c.html.

139   "[Clooney] said that Bush . . .": quoted in BBC News, "Clooney in Anti-War Protest," January 20, 2003, http://news.bbc.co.uk/2/hi/ entertainment/2677881.stm.

139   "Shortly after George . . .": Barbra Streisand, "Barbra Streisand at the Democratic Congressional Campaign Committee Gala," September 29, 2002, http://www.barbrastreisand.com/news_ statements.html.

139   "If the *NY Post* . . .": Barbra Streisand, Truth Alert, http://www. barbrastreisand.com/news_truth.html.

139   "She used a speech . . .": John Schwartz, "Brawling with Barbra Online," *New York Times,* October 6, 2002.

140   "She identified Saddam . . .": Jennifer Harper, "Hollywood Takes
       on White House," *Washington Times*, October 19, 2002, http://
       www.washtimes.com/national/20021019-86368341.htm.

140   "spelling-bee champion . . .": Tammy Bruce, "Funny Lady,"
       NewsMax.com, October 2, 2002, http://www.papillonsartpalace.
       com/funnylad.htm.

140   "called her from Air Force One . . .": Mark Steyn, "Let Slip the
       Babs of War," *Daily Telegraph*, October 19, 2002, http://www.
       telegraph.co.uk/opinion/main.jhtml?xml=/opinion/2002/10/19/do1
       902.xml.

140   "Just so you . . .": quoted in Associated Press, "Critical Comments
       About U.S. President Gets Dixie Chicks Singer into Trouble,"
       March 19, 2003.

141   "For me as an American . . .": quoted on Ananova.com, "Hoff-
       man Blasts Bush's War Plans."

141   "hijacked our pain . . .": quoted in Alice Klein, "U.S. Takes on
       Peace," NowToronto.com, http://www.nowtoronto.com/issues/
       2002-10-31/news_feature.php (2/25/03).

142   "Wisdom comes . . .": Aeschylus, *The Orestia*, translated by
       Robert Fagles (New York: Penguin, 1975), lines 179–184.

142   "the highest leaders . . .": A Statement of Conscience: Not in
       Our Name, http://notinourname.net/statement_conscience.html
       (2/24/03).

142   "war without limit": Ibid.

142   "Those in public . . .": Sontag in Hertzberg et al, "Tuesday and
       After," *The New Yorker*.

144   "The difference between . . .": Nicholas Lemann, "The War on
       What? The White House and the Debate About Whom to Fight
       Next," *The New Yorker*, September 9, 2002.

145   "undermining . . . because Kofi Annan . . .": quoted in William F.
       Buckley Jr., "Mandela's Contribution," nationalreviewonline,
       January 31, 2003, http://www.nationalreview.com/buckley/
       buckley013103.asp.

145   "house slave": quoted in CNN.com, "Belafonte Won't Back Down
       From Powell Slave Reference," October 15, 2002, http://www.cnn.
       com/2002/US/10/15/belafonte.powell.

145   "highlight . . . areas of concern . . .": quoted in James Baehr,
       "Speakers in Review: Al Sharpton," *Dartmouth Review*, http://
       www.dartreview.com/archives/000360.php (2/7/03).

145 "I ... think that there is a strong streak ...": quoted in "Ed Asner Blames America First & Laughs-Off Bush's Vision," Media Research Center: Cyber Alerts, November 25, 2002, http://www. mediaresearch.org/cyberalerts/2002/cyb20021125.asp#4.

145 "after September 11 ...": Senator Russ Feingold, "Why I Opposed the Anti-Terrorism Bill," Counterpunch.org, edited by Alexander Cockburn and Jeffrey St. Clair, October 26, 2001.

146 "Indeed, the government ...": Emily Bazar, "INS Policy Sparks Furor," *Sacramento Bee*, December 21, 2002, http://www.sacbee. com/content/news/california/story/5667369p-6641988c.html (2/13/03).

147– "The movement to restore ...": Lynda Gorov, "Hollywood Tak-
148 ing Up Cause of Afghan Women," *Boston Globe*, March 18, 1999.

148 "President Bush is ...": http://www.bonnieraitt.com/activism.php (2/7/03).

148 "The basic reason ...": Noam Chomsky, "Mirror Crack'd—The Effect of 9/11 Was Much the Same as the Cause; a Morally-Void, Global Soliloquy of Power," OutlookIndia.com, September 16, 2002 (2/24/03).

148 "Not in our name will you wage ...": A Statement of Conscience; Not in Our Name.

148 "knows the real reason we are invading ...": Barbra Streisand, "My Thoughts Today," February 20, 2003, http://www. barbrastreisand.com/news_statements.html.

148 "Nelson Mandela has also joined ...": William F. Buckley Jr., "Mandela's Contribution," nationalreviewonline, January 31, 2003.

149 "oil gluttony is ...": Barbara Kingsolver, "No Glory in Unjust War on the Weak," *Los Angeles Times*, October 14, 2001, p. M1.

149 "Iraq produces 2 million barrels ... The total global ...": *The Guardian*, "Facts About Iraq's Oil Production," http://www. guardian.co.uk/uslatest/story/0,1282,-2297850,00.html, January 6, 2003 (2/25/03).

150 "huge karmic retributions ...": quoted in Andrew Sullivan, "Idiocy of the Week," Salon.com, http://www.salon.com/opinion/ sullivan/2003/01/15/crow/, January 15, 2003 (2/24/03).

150 "the war against terrorism ...": Mirror.co.uk, "Woody's On Side," http://www.mirror.co.uk/news/allnews/page.cfm?objectid= 12103357&method=full&siteid=50143.

150   "answered one terrorist act . . .": Kingsolver, "No Glory in Unjust War on the Weak."

151   "September 11 was not all that different . . .": Adlai E. Stevenson III, "Different Man, Different Moment," *New York Times,* February 7, 2003.

151   "compassion and understanding": quoted in MSNBC.com, "McCartney Caps Huge NY Benefit," http://www.msnbc.com/news/645450.asp, October 21, 2001 (2/24/03).

151   "we must be mindful . . .": Gerda Lerner, "Alternatives to War Will Work Best in Long Run," Madison *Capital Times,* October 1, 2001.

152   "I feel like I'm standing . . .": Kingsolver, "No Glory in Unjust War on the Weak."

152   "in a war on Afghanistan . . .": Alice Walker, *Sent by Earth: A Message from the Grandmother Spirit After the Attacks on the World Trade Center and the Pentagon* (New York: Seven Stories Press, December 2001).

152   "We must insist . . .": quoted in John Perazzo, "The Anti-American: Medea Benjamin," FrontPageMagazine.com, November 15, 2002, http://www.frontpagemag.com/Articles/Printable.asp?ID= 4631 (2/7/03).

153   "While the Pentagon makes no estimate . . .": Carl Conetta, "Strange Victory: A Critical Appraisal of Operation Enduring Freedom and the Afghanistan War," Project on Defense Alternatives Research Monograph #6, January 30, 2002, http://www.comw.org/pda/0201strangevic.html (2/13/03).

153   "Not in our name will you invade . . .": Notinourname.net, "War Without End? Not in Our Name—The Pledge of Resistance" (2/7/03).

153   "More than a decade . . .": http://www.bonnieraitt.com/activism.php (2/7/03).

154   "Everybody is telling me to bomb . . .": Woody Harrelson, "I'm an American Tired of American Lies," *The Guardian,* October 17, 2002, http://www.guardian.co.uk/g2/story/0,3604,813189,00.html (2/6/03).

155   "We're picking on people . . .": quoted in Media Research Center—Cyber Alert January 23, 2003, http://www.mediaresearch.org/cyberalerts/2003/cyb20030123.asp#7.

155   "Quite probably the worst thing . . .": quoted in Shapiro, "War or
        Peace? Local notables Take Their Stand."

156   "the Israeli-Palestinian conflict . . .": Stevenson, "Different Man,
        Different Moment."

156   "brutal repercussions": "A Statement of Conscience: Not in
        Our Name," http://notinourname.net/statement_conscience.html
        (2/24/03).

156   "More than seven hundred Israelis . . .": Council on Foreign Rela-
        tions, "Flashpoint: Israeli-Palestinian Conflict," http://www.
        terrorismanswers.com/policy/israel.html.

157   "supporting terror": Stevenson, "Different Man, Different
        Moment."

159   "Whether made by . . .": Ibid.

159   "Unless we attack . . .": Lerner, "Alternatives to War Will Work
        Best in Long Run."

159   "the son of a billionaire . . .": Adam Cohen, Michael Weisskopf,
        Adam Zagorin and Bruce Crumley, "How Bin Laden Funds His
        Network," Time, October 1, 2001, vol. 158, issue 15, p. 63.

160   "the U.N. Relief and Works Agency . . . will spend $296
        million . . .": United Nations Relief and Works Agency, Esta-
        blishment of UNRWA, http://www.un.org/unrwa/about/index.html
        (2/12/03).

160   "Indeed, Forbes recently . . .": "Kings, Queens, and Despots,"
        Forbes.com, February 27, 2003, http://www.forbes.com/work/
        compensation/2003/02/24/0224kings.html.

160   "You can't defend America . . .": Remarks by Barbra Streisand,
        Rainbow/PUSH Coalition Fourth Annual Awards Dinner,
        December 11, 2001, http://www.barbrastreisand.com/news_
        statements10.html.

160   "I find the erosion . . .": Barbra Streisand, remarks at the Demo-
        cratic Congressional Campaign Committee Gala, September 29,
        2002.

160   "those to whom the basic rights . . .": A Statement of Conscience:
        Not in Our Name (2/24/03).

160   "Not in our . . .": Notinourname.net, "War Without End? Not in
        Our Name—The Pledge of Resistance," http://www.notinourname.
        net/ (2/7/03).

161 "We haven't forsaken . . .": quoted in Evan P. Schultz, "Memo to Ashcroft: Read 'Marbury'," LegalTimes.com, October 25, 2001.

161 "government officials, so long as their primary purpose . . .": Ibid.

162 "Use . . . information . . .": Feingold, "Why I Opposed . . ."

163 "we suspect that these communications . . .": Testimony of Attorney General John Ashcroft, Senate Committee on the Judiciary, December 6, 2001, http://www.usdoj.gov/ag/testimony/2001/1206 transcriptsenatejudiciarycommittee.htm.

164 "Ashcroft's insistence . . .": CNN.com, "Attorney Among 4 Accused of Supporting Terrorism," April 10, 2002, http://www.cnn.com/2002/LAW/04/09/inv.terror.indictment/index.html (2/12/03).

164 "We have asked . . .": Testimony of Attorney General John Ashcroft, Senate Committee on the Judiciary, December 6, 2001, http://www.usdoj.gov/ag/testimony/2001/1206transcriptsenate judiciarycommittee.htm.

164 "six hundred Taliban and al Qaeda fighters . . .": BBC News, "Amnesty Pleads for Guantanamo Inmates," January 11, 2003, http://news.bbc.co.uk/2/hi/americas/2648241.stm.

165 "legal black hole": quoted in Ibid.

165 "There is no . . .": quoted in Katty Kay, "No fast track at Guantanamo Bay," BBC News, January 11, 2003, http://news.bbc.co.uk/2/hi/americas/2648547.stm.

165 "There are reports . . .": Ibid.

165 "After 9/11, a massive, worldwide dragnet . . .": Celina B. Realuyo, Policy Advisor, U.S. Department of State, Counterterrorism Office—Remarks to Western Union International Compliance Conference, October 25, 2002.

165 "twelve hundred Middle Eastern . . .": Human Rights Watch, "United States: Abuses Plague Sept. 11 Investigation—Checks on Government Authority Should Be Restored," August 15, 2002, http://www.hrw.org/press/2002/08/usdetainess081502.htm (2/12/03).

165 "Most were deported . . .": Human Rights Watch, "PRESUMPTION OF GUILT: Human Rights Abuses of Post-September 11 Detainees," August, 2002, http://www.hrw.org/reports/2002/us911/USA0802.htm#P86_1667.

166 "would enable a team . . .": Adam Clymer, "Threats and Responses: Electronic Surveillance; Congress Agrees to Bar Pentagon from Terror Watch of Americans," *New York Times,* February 12, 2003.

166   "We [the United States] have been . . .": Joe Kovacs, "Rush Lim-
        baugh: Bill Maher 'Was Right,' " WorldNet Daily, September,
        2001, http:// www.worldnetdaily.com/news/article.asp?ARTICLE_
        ID=25267.

166   "If the word . . .": Sontag in Hertzberg et al, "Tuesday and After."

167   "Indeed, the total of American combat deaths . . .": Interview with
        Lt. Jim Casella of Pentagon Information Bureau, January 12, 2003.

168   "scaredy cats": quoted in Catherine Donaldson-Evans, "Brits
        Bristling at Michael Moore's One-Man Act," FOXnews.com, Jan-
        uary 8, 2003, http://www.foxnews.com/story/0,2933,74883,00.
        html.

168   "If they [the Bush administration] . . .": quoted in Ananova.com,
        "Hoffman Blasts Bush's War Plans."

Chapter 5

170   "fifty-six thousand six hundred eighty-one American troops . . .":
        According to the American Battle Monuments Commission, there
        are 26,255 Americans from World War I buried in four French
        cemeteries, and 30,426 Americans from World War II buried in six
        cemeteries in France.

173   "Ever since France relinquished . . .": www.islamicweb.com,
        hosted by WebSolution, updated December 28, 2000.

173   "dismissed the Holocaust . . .": Emelia Sithole and Glaieul Mam-
        aghani, "Le Pen Triumph Spreads Fear Among France's Jews,"
        Reuters, April 22, 2002.

173   "officially admitted that France . . .": Ibid.

173   "their authorities helped the Nazis . . .": Joseph Fareh, "Anti-
        Semitism Rears Ugly Head in Europe," http://wordofmessiah.org/
        france_antisemitism.htm, January 13, 2003.

174   "a thirty-four-year-old . . .": Jeff Jaesly, "The Canary in Europe's
        Mine," Boston Globe, April 28, 2002.

174   "French schools recorded 455 . . .": Marc Perelman, "French Min-
        ister Unveils Plan to Fight Anti-Semitism," Forward, March 7,
        2003.

174   "A synagogue in Marseilles . . .": "Synagogues in France, Belgium
        Set on Fire in Spate of Attacks," Washington Post, April 2, 2002,
        p. 16.

174   "A car bomb . . . in Montpellier . . . a synagogue in Stras-
        bourg . . . a Jewish Sports Club . . . in Bondy . . . the bus that
        takes . . .": Jaesly, "The Canary in Europe's Mine."

174 "In total, French police...": The Steven Roth Institute for the Study of Anti-Semitism and Racism, "Anti-Semitism World-Wide, 2001/2002: France" http://www.tau.ac.il/anti-semitism/asm2001:2/france.htm.

174 *"Dreaming of Palestine...":* "The French Intifada," *FrontPage Magazine,* Frontpagemag.com, January 28, 2003.

174 "The governing body of the Pierre and Marie Curie campus...": Ibid.

175 "Had they [the French government] taken measures...": Sithole and Mamaghani, "Le Pen Triumph Spreads Fear Among France's Jews."

175 "facing a dangerous wave...": quoted in "Are the French Really Anti-Semitic," *Le Monde Diplomatique,* December 2002, http://mondediplo.com/2002/12/14antisemitism.

175 "the West's worst country...": quoted in Ibid.

176 "Modern Europe's visceral...": David Gelernter, "The Roots of European Appeasement," *Weekly Standard,* September 23, 2002.

178 "peace in our time...": Speech by Neville Chamberlain, September 30, 1938, http://www.lib.byu.edu/~rdh/eurodocs/uk/peace.html (the phrase is also cited frequently as "peace for our time" in historical accounts).

178–
179 "the victorious allies...": Gelernter, "The Roots of European Appeasement."

179 "I hate England...": quoted in William L. Shirer, *The Collapse of the Third Republic* (New York: Simon & Schuster, 1969), p. 250.

180 "even though the captives themselves...": Rev. Fr. Charles T. Brusca, "Psychological Responses to Terrorism," http://www.yahoodi.com/peace/stockholm.html.

181 *"L'appétit vient en mangeant...":* Stephen R. Rock, *Appeasement in International Politics,* University of Kentucky Press, p. 2.

182 "is a large, friendly dog...": Arnold Toynbee, quoted on http://www.quoteproject.com/subject.asp?subject=56.

182 "determined, after the humiliating defeat...": Richard Bernstein, *Fragile Glory: A Portrait of France and the French* (New York: Knopf, 1990), p. 133.

183 "an imperative of... was fated to be the...": Ibid., pp. 134–135.

183 "the weak grow strong..." Henry Kissinger, *Years of Upheaval,* (Little, Brown: Boston, 1987), p. 173.

183 "has many sides...": Bernstein, *Fragile Glory,* p. 135.

185 "But the war with Iran . . .": Brent Sadler, "The Unfinished War: A Decade Since Desert Storm," CNN, January 16, 2001.

185 "while mobilizing . . ." Jim Bitterman, "France: The Ambiguous Ally," CNN, January 17, 2001.

185 "Until the last . . .": quoted in Ibid.

185 "It was only after the French Embassy . . ." *Middle East Digest*, December 1996.

185 "France, with 4.5 million . . .": Jim Hoagland, "Europeans Still Firm in Opposing Saddam; Poll Indicates Support for Military Action," *Washington Post*, October 25, 1990, p. A31.

185 "There are those who fear . . .": quoted in Ibid.

186 "at the [Kuwait] border . . .": Ibid.

186 "On March 31, 1995 . . .": Lee M. Katz, "The Joke's on Us, Amid French-twisted Election Rhetoric," *USA Today*, March 31, 1995, p. 13A.

186 "The *London Mail* said that . . .": *Middle East Digest*, December 1996.

186 "The *Jerusalem Post* quoted . . .": quoted in Ibid.

186 "According to Physicians for Human Rights . . .": Sadler, "The Unfinished War."

186 "In August of 1995, Saddam . . .": Ibid.

187 "We were ordered to hide everything . . .": quoted in Ibid.

187 "a policy of getting out of the crisis . . .": quoted in Ibid.

187 "These deals, however . . .": Roula Khalaf, "Death by Sanctions: Iraq, Did Washington Spike the Sector's Recovery Programme?" *Financial Times*, April 15, 1999, p. 07.

188 "a small carrot": John Harris, "Clinton's 2nd-Term Team Undergoes an Acute Test; Search for Diplomatic Solution on Iraq Scrutinized in Light of Strong Public Stance," *Washington Post*, November 20, 1997, p. A36.

188 "French diplomats describe . . .": Michael Littlejohns, "France Launches Drive to End Oil Embargo on Baghdad," *Financial Times*, January 14, 1999, p. 06.

188 "monitoring . . .": Ibid.

188 "a first step . . . sees a need for a balanced . . .": Ibid.

188   "oil embargo on Iraq . . .": Nicole Winfield, "France Presents Iraq Embargo Plan," Associated Press, January 13, 1999.

189   "an ongoing monitoring system . . .": Ibid.

189   "immediately shift UNSCOM's . . .": Ibid.

189   "insisting that it failed . . .": Stephen Fidler and Roula Khalaf, "No Policy Change, Just a Shot Across Saddam's Bows," *Financial Times,* February 17, 2001, p. 6.

190   Fox News/Opinion Dynamics Survey, March 25–26, 2003.

Chapter 6

196   "the company's stock . . .": Daniel Kadlec, "Enron: Who's Accountable?" *Time,* January 13, 2002.

196   "or about $1.2 billion . . .": Leslie Wayne, "Before Debacle, Enron Insiders Cashed in $1.1 Billion in Shares," *New York Times,* January 13, 2002.

196   "As Enron stock climbed . . . sold Enron stock . . .": Ibid.

196   "Roy was one of the unlucky ones . . .": William Lerach and Al Meyerhoff, "Why Insiders Get Rich and the Little Guy Loses," www.enronfraud.com/insidervslittl.html, January 20, 2002.

197   "The *Chicago Tribune* . . .": Sam Roe, Cam Simpson, and Andrew Martin, "Trail of Complaints About Andersen," *Chicago Tribune,* January 27, 2002.

198   "The story began . . .": Central Bank of Denver, N.A., v. First Interstate Bank of Denver, N.A., 511 U.S. 164, 114 S.Ct. 1439, 128 L. Ed. 2d 119, 62 USLW 4230, Fed. Sec. L. Rep P 98,178 (U.S. Colo. Apr. 19, 1994), (NO. 92–854), http://supct.law.cornell.edu/supct/html/92-854.ZO.html.

198   "aiding and abetting": Ibid.

198   "In *hundreds* of judicial and administrative proceedings . . .": *New England Law Review,* Winter 1996, http://www.nesl.edu/lawrev/vol30/vol30-2TOC.htm.

198   "deters secondary actors . . .": "Securities Law Litigation Following 'Central Bank,' " *New York Law Journal,* New York Publishing Company, November 17, 1994.

198   "common sense legal reform": "The Common Sense Legal Reform Act," HR 10, http://www.house.gov/house/Contract/legalrefd.txt.

199 "the *New York Times* reported . . .": George Judson, "Accountants to Pay $10 Million to Victims of Real Estate Fraud," *New York Times*, April 24, 1996.

199 "the investors lost tens of millions": Judson, "Accountants to Pay $10 Million to Victims of Real Estate Fraud."

199 "$90 million": Mark Pazniokas, "Judge Approves Settlement in Landmark Case," *Hartford Courant*, June 11, 1997.

200 "as U.S. Attorney . . .": Mark Pazniokas, "A Dream Come True," *Hartford Couraut*, October 7, 1997.

200 "original draft": David Lightman, "Dodd Faces Consumer Critics on Securities Litigation Bill," *Hartford Courant*, March 15, 1994.

200 "put cases currently . . .": Ibid.

200 "Ralph Nader called Dodd's legislation . . .": David Lightman, "Dodd Defends Himself Against Accusations of Consumer Groups," *Hartford Courant*, June 28, 1995.

200 "would set a . . .": Richard Keil, "Accountants Shielded from Lawsuits; Legislation Crafted by Dodd," *Legal Intelligencer*, March 24, 1994.

200 "Michael Calabrese, executive . . .": David Lightman, "Dodd Defends Himself Against Accusations of Consumer Groups," *Hartford Courant*, June 28, 1995.

200 "Dodd's bill also . . .": Lightman, "Dodd Faces Consumer Critics on Securities Litigation Bill."

201 "safe harbor provision . . .": "Private Securities Litigation Reform of 1995: Utilizing the Safe Harbor," www.perkinscoie.com/resource/business/securities.htm.

201 "requires that a victim's complaint . . .": Consumer Federation of America, "In the Wake of Enron, Consumer Groups Urge Congress to Restore Investor Confidence by Updating 1996 Private Securities Litigation Law," press release, February 15, 2002, http://www.consumersunion.org/finance/securdc202.htm.

201 "a Catch-22 . . .": quoted in Ibid.

201 "It allows for an award . . .": Gary Lawson and Walter Olson, "Civil Justice Memo 16—Caveat Auditor: The Rise of Accountants' Liability," Manhattan Institute for Policy Research, May 1989, http://www.manhattan-institute.org/html/cjm_16.htm.

201 "changes were a . . .": William Lerach, "The Private Securities Litigation Reform Act of 1995—27 Months Later," Securities Class

Action Litigation Under the Private Securities Litigation Reform Act's Brave New World," *Washington University Law Quarterly,* vol. 76, no. 2, Summer 1998.

201 "$54,843 from Arthur Andersen": "Top Senate Recipients of Anderson PAC & Individual Contributions, 1989–2001," The Center for Responsive Politics at opensecrets.org, based on FEC data, http://www.opensecrets.org/news/enron/andersen_senate_top.asp.

201 "$37,750": "Alert/Computer Industry Campaign Contributions," October 2, 1998, The Center for Responsive Politics at opensecrets.org, based on FEC data, http://www.opensecrets.org/alerts/v4/alrtv4n34.asp.

202 "perhaps the accounting . . .": Stephen Labaton, "Enron Enablers," *New York Times,* February 2, 2002.

202 "Overall, during the 1995–1996 campaign cycle . . .": "Prevent Financial Fraud: Repeal the Accountant Immunity Act—The 1995 Private Securites Litigation Reform Act (PSLRA)," Enron Watchdog, www.enronwatchdog.org/topreforms/topreforms5.html.

204 "double crosser": David Maraniss, *First in His Class: A Biography of Bill Clinton* (New York: Touchstone, 1996), p. 409.

204 "I am not willing to sign . . .": Bill Clinton's Veto Message Re Private Securities Litigation Reform Act, December 20, 1995.

204 "Twenty Democrats . . . in the Senate": "On Overriding the Veto (H.R. 1058 passage over veto), U.S. Senate Roll Call Votes, 104th Congress–1st Session, December 22, 1995, http://www.senate.gov/legislative/LIS/roll_call_lists/roll_call_vote_cfm.cfm?congress=104&session=1&vote=00612.

205 "eighty-nine Democratic congressmen": Securities Litigation Reform Act, Passage Objections of the President Notwithstanding, December 20, 1995, Vote http://clerkweb.house.gov/cgi-bin/vote.exe?year=1995&rollnumber=870.

205 "Chris Dodd—Here he is . . .": *Frontline,* "Bigger Than Enron: Congress and the Accounting Wars," The Battle Over Tort Reform, www.pbs.org/wgbh/pages/frontline/shows/regulation/congress/.

205 "cut off investors . . . sure some of . . .": Ralph Z. Hallow, "GOP Challenges Clinton on Oversight," *Washington Times,* July 26, 2002, http://www.washtimes.com/national/20020726-53936167.htm.

205 "At the heart . . .": Kadlec, "Enron: Who's Accountable?"

206   "the driving force": Kurt Eichenwald, "Andersen Trial Yields Evidence in Enron's Fall," *New York Times,* June 17, 2002.

207   "In 2001, for . . .": Kadlec, "Enron: Who's Accountable?"

207   "one of the most aggressive . . .": Morton Kondracke, "Assessing the Blame," *Washington Times,* January 29, 2002.

207   "accounting firms are . . .": Associated Press, "SEC Proposes New Auditor Rules," June 28, 2000.

207   "feverishly . . . crisscrossing the country . . .": Sandra Sugawara, "Accounting Role in the Balance; Levitt Lobbying to Save Proposed Curbs on Firms' Consulting Work," *Washington Post,* October 24, 2000.

207   "sparked a firestorm of protests": Ibid.

207   *"USA Today* reported . . .": Greg Farrell, "Congress Could've Seen It Coming," USAToday.com, January 17, 2002, http://www.usatoday.com/money/covers/2002-01-17-bcovthu.htm.

207   "Tauzin, Chairman of . . .": Ibid.

207   "no evidence": Ibid.

207   "Schumer had taken . . .": Ibid.

207   "was almost certainly . . .": Kondracke, "Assessing the Blame."

207   "After the Enron . . .": Farrell, "Congress Could've Seen It Coming."

208   "He says that . . .": Ken Fireman, "How Andersen Blocked SEC Rule Change," Newsday.com, January 17, 2002, http://www.newsday.com/business/nationworld/sns-enron-secrules-nyn.story.

208   "cooking the books": Farrell, "Congress Could've Seen It Coming."

208   "force dramatic changes . . . to pay increased . . .": Associated Press, "Summers Warns Lawmakers Not to Intervene in SEC Proposal," October 6, 2000.

208   "Gramm's wife, Wendy . . .": *Frontline,* "So You Want to Buy a President?: Wendy Gramm" http://www.pbs.org/wgbh/pages/frontline/president/players/gramm.html.

208   "She sold all . . .": Leslie Wayne, "Before Debacle, Enron Insiders Cashed in $1.1 Billion in Shares," *New York Times,* January 13, 2002.

208   "helped broker a ...": quoted in Opinion, "Pernicious Atmosphere—Is Dodd a Figure in Enron Swamp?" Associated Press/*Winchester Star*, February 13, 2002.

209   "The 1995 securities law ...": James E. Day, "The Securities Litigation Uniform Standards Act of 1998: Solution or Stop Gap Measure," Prentice Hall Law and Business, February 1999, and *Tech Law Journal*, "House Passes Securities Class Action Litigation Reform Bill," July 23, 1998.

209   "by placing procedural ...": Ibid.

209   "The solution to ...": Ibid.

209   "an untrue statement ...": Ibid.

209   "designed to decrease ...": "House Passes Securities Class Action Litigation Reform Bill," *Tech Law Journal*.

210   "our thriving high technology ...": quoted in Ibid.

210   "If we pass ...": quoted in Ibid.

211   "Accountants would be ...": USAToday.com, "Sarbanes-Oxley Act's Progress," December 26, 2002, http://www.usatoday.com/money/companies/regulation/2002-12-26-sarbanes_x.htm.

211   "Around the jail ...": Howard Fineman; Michael Isikoff; Tamara Lipper; Julie Scelfo, "Laying Down the Law," *Newsweek*, August 5, 2002.

212   "Wait'll you see what's next ...": Ibid.

212   "The Rigas arrests ...": Ibid.

212   "The final law ...": quoted in Godfrey and Kahn, "President Bush Signs Corporate Reform Legislation," August 1, 2002, http://www.gklaw.com/InfoResources/ircorporategovernance.pdf.

214   "rigged for the Wall Street houses": Author interview with Robert Weiss, February 20, 2003.

Chapter 7

216   "Even the Communist Party ...": Richard E. Cohen, "When Campaigns Are Cakewalks," *National Journal*, March 16, 2002.

216   "With Hispanic voters accounting ...": David E. Rosenbaum, "Campaign Season," *New York Times*, October 3, 2002 p. A 24.

216   "Governor Gray Davis and the Democratic Legislature ...": Ibid.

217 "What they have . . .": quoted in Ibid.

217 "The five incumbent California congressmen . . .": Author survey based on information from: CNN.com Election Results 2000, http://www.cnn.com/ELECTION/200/results/house/index.html (1/16/02); CNN.com Election Results 2002, http://www.cnn.com/ELECTION/2002/pages/house/index.html (1/16/02).

217 "California is the poster child . . .": *Monopoly Politics 2002: How "No Choice" Elections Rule in a Competitive House*, The Center for Voting and Democracy, September 30, 2002, http://www.fairvote.org/2002/mp2002.htm (12/17/02).

217 "if the average . . .": Alison Mitchell, "Redistricting 2002 Produces No Great Shake-Ups," *New York Times*, March 13, 2001.

217 "California's redistricting plan . . . My staff and I . . .": Ibid.

217 "skinny as a snake . . . slightly Republican leaning . . .": Adam Graham-Silverman, "Redistricting Gurus Protect Incumbents," *CQ Weekly*, vol. 60, issue 20, April, 18, 2002, p. 1279.

218 "Rather than press for greater gains . . .": Ibid.

218 "To the dismay of some . . .": Mitchell, "Redistricting 2002 Produces No Great Shake-Ups."

218 "Florida may have trouble . . .": Jonathan Allen, "Fla. Remap Gives Most Incumbents Home-Field Edge," WashingtonPost.com, *CQ Monitor*, August 30, 2002, http://www.washingtonpost.com/ac2/wpdyn?pagename=article&node=politics/elections/redistricting&contentId=A16972-2002Aug30&notFound=true, (2/19/03).

219 "preclearance . . . undermine a main . . .": David E. Rosenbaum, "Justice Dept. Accused of Politics in Redistricting," *New York Times*, May 31, 2002.

219 "action by a federal court . . .": Adam Clymer and Paul Zielbauer, "In 4 Races, an Incumbent Is Guaranteed to Lose," *New York Times*, October 27, 2002, Section 1, p. 3.

220 "But Junior blew it . . . four arrests . . .": Sylvia Cooper, "Trials, Triumphs Through Life Lead Walker to New Challenge," *Augusta Chronicle* Online, http://augustachronicle.com/stories/102002/met_walker_proflie.shtml (2/19/03).

220 "six districts with [at least] 54 percent . . .": Dave Hamrick, "Local GOPers Condemn 'Splatter Art' Redistricting of House, Senate," *The Citizen Online*, December 5, 2001 (1/16/03).

220 "There's no question . . .": David M. Halbfinger and Jim Yardley, "Vote Solidifies Shift of South to the G.O.P.," *New York Times,* November 7, 2002, p. A1.

221 "The Democrats have controlled . . .": quoted in Dave Hamrick, "Local GOPers Condemn 'Splatter Art' Redistricting of House, Senate," *The Citizen Online,* December 5, 2001 (1/16/03).

221 "despite the legislature . . .": Noelle Straub, "Georgia Leads the Field in Upset Races," The Hill.com, January 16, 2003, http://www.thehill.com/news/110602/georgia.aspx (1/16/03).

221 "Democrats . . . are in a hell of a box . . .": Richard E. Cohen, "Politics: Redrawing the House," *National Journal,* April 7, 2001.

221 "Accused of helping a developer buddy . . .": Michelle Malkin, "Democrats' Bad Boy," *Washington Times* Online, July 9, 2002, http://www.washtimes.com/commentary/20020709-45588744.htm (2/20/03).

222 "I'll break your nose": Ibid.

222 "after getting that . . .": Ibid.

222 "has been entangled . . .": Nichole M. Christian, John H. Cushman Jr., Sherri Day, Sam Dillon, Neil A. Lewis, Robert Pear, Terry Pristin, Philip Shenon, Jacques Steinberg, and Leslie Wayne, "The 2002 Elections: South; Virginia," *New York Times,* November 6, 2002, p. B6.

222 "deal with state Republicans": Ibid.

222 "despite the increasing majority . . .": Associated Press, "Court Rejects Challenge to Mich. Redistricting," *Washington Post,* May 25, 2002, p. A12.

223 "Democrats controlled about . . .": Norman J. Ornstein, "Why Close Races Ruin Politics," *New York Times,* November 4, 2002.

223 "the rest were split": Actual count: Democrats 18, Republicans 17, split 14.

223 "Most polls show[ed] . . .": Ornstein, "Why Close Races Ruin Politics."

223 "Twenty of the thirty-six governorships . . .": David J. Garrow, "Ruining the House," *New York Times,* November 13, 2002, section A, p. 29.

224 "Ninety-six percent . . .": Author survey based on information from CNN.com, "Election Results 2000."

224   "90 percent of . . .": David J. Garrow, "Ruining the House," *New York Times,* November 13, 2002, section A , p. 29.

224   "the House of Representatives . . .": Clinton Rossiter, ed., *The Federalist Papers* (New York: Mentor, 1961), p. 358.

224   "immediate representatives of . . .": Ibid.

225   "Columnist Bill Kristol cites this cyber-revolution . . .": Conversation with author.

226   "eight states gained . . .": Table 384, U.S. Census Bureau, Statistical Abstracts of the United States, 121st edition; Washington, DC, 2001.

227   "In most . . . states . . .": Gregory L. Giroux, "Redistricting Helped GOP," *CQ Weekly,* vol. 60, issue 43, November 9, 2002, p. 2934.

228   "insider trading, just . . .": op-ed, Steven Hill and Rob Richie, "Why 95 Percent of House Incumbents Won," *The Hill,* http://www.thehill.com/op_ed/110602_hillrichie.aspx (1/16/03).

228   "In 2002, only sixteen incumbent . . .": Gregory L. Giroux, "Redistricting Helped GOP," *CQ Weekly,* vol. 60, issue 43, November 9, 2002, p. 2934.

229   "Not only did virtually every House incumbent . . .": Author survey based on information from CNN.com Election Results 2002, CNN.com, http://www.cnn.com/ELECTION/2002/pages/house/index.html (1/16/02).

229   "Thirty-three House members . . .": Giroux, "Redistricting Helped GOP."

229   "Winning Vote Share . . .": Author survey based on information from: Ibid and CNN.com Election Results 2002, http://www.cnn.com/ELECTION/2002/pages/house/index.html (1/16/02).

229   "congressmen who won only narrowly in 2000 . . .": Center for Voting and Democracy, "Factsheet: Incumbents Fared Well in Redistricting," October, 2002, http://www.fairvote.org/2002/incumbents.html (12/17/02).

229   "Eighteen of them . . .": Author survey based on information from CNN.com Election Results 2000, http://www.cnn.com/ELECTION/200/results/house/index.html (1/16/02). CNN.com Election Results 2002, http://www.cnn.com/ELECTION/2002/pages/house/index.html (1/16/02).

229   "had their districts made safer . . .": The Center for Voting and Democracy, "Factsheet: Incumbents Fared Well in Redistricting,"

October 2002, http://www.fairvote.org/2002/incumbents.htm (12/17/02).

230 "Indeed, the *Cook* . . .": Mitchell, "Redistricting 2002 Produces No Great Shake-Ups."

230 "Perhaps most alarming . . .": "Monopoly Politics 2002: How 'No Choice' Elections Rule in a Competitive House," Center for Voting and Democracy, September 30, 2002, http://www.fairvote.org/2002/mp2002.htm (12/17/02).

230 "most voters have . . .": Steven Hill and Rob Richie, "Get Your Election Results Here: 99.8% Accurate," Center for Voting and Democracy, November 3, 2002, http://www.fairvote.org/op_eds/houston110302.htm (2/19/03).

231 "Nationally, only 39 percent . . .": Peter Beinart, "Outer Limits," *New Republic,* November 25, 2002, vol. 227, issue 22, p. 6.

231 "I always voted . . .": Gilbert and Sullivan, *H.M.S. Pinafore,* from "The Legal Songs of Gilbert and Sullivan," *Lexscripta,* http://www.lexscripta.com/links/cultural/pinafore.html (2/19/03).

232 "Judging from last . . .": David J. Garrow, "Ruining the House," *New York Times,* November 13, 2002, section A, p. 29.

233 "four, essential, measurable criteria . . .": Public Interest Guide to Redistricting, "Iowa's Redistricting Information," http://www.fairvote.org/redistricting/reports/remanual/ia.htm (2/19/03).

233 "We don't even . . .": Gregory L. Giroux, "In Iowa, Redistricting Without Rancor," *CQ Weekly,* April 7, 2001, vol. 59, issue 14, p. 792.

233 "We're just all calmly sitting back . . .": quoted in Ibid.

233 "It cuts out . . .": quoted in Ibid.

233 "Republicans cared most about the partisan advantage . . .": Adam Clymer, "Why Iowa Has So Many Hot Seats," *New York Times,* October 27, 2002, p. 5.

233 "in one respect . . .": "In Praise of Iowa," *The Economist,* October 19, 2002, vol. 365, issue 8295, p. 2.

233 "Congressman Jim Leach won with 52 percent . . .": CNN.com Election Results 2002, http://www.cnn.com/ELECTION/2002/pages/house/index.html (1/16/02).

234 "Voters in Arizona . . .": "In Praise of Iowa," *The Economist.*

Chapter 8

236   "Because of their guts and vision . . .": American Lung Association, "State of Tobacco Control 2002," January 7, 2003, http:// www.lung usa.org/press/tobacco/tobacco_010703_d.html (1/21/03).

236   "including almost $9 billion . . .": State Tobacco Settlement, Campaign for Tobacco-Free Kids, http://tobaccofreekids.org/reports/ settlements/ (2/15/03).

237   "Spurred by a . . .": "Big Tobacco Wins, Kids and Taxpayers Lose as California Budget Slashes Tobacco Prevention and Fails to Increase the Cigarette Tax," Campaign for Tobacco-Free Kids, http:// tobaccofreekids.org/Script/DisplayPressRelease.php3?Display=535 (2/4/03).

237   "Tobacco causes approximately . . .": National Cancer Policy Board, Institute of Medicine and National Research Council, "State Programs Can Reduce Tobacco Use" (National Academies Press, 2000), http://books.nap.edu/html/state_tobacco/, p. 2 (1/28/03).

237   "about one-fifth of all fatalities . . .": (430,000/2,388,070), Johns Hopkins School of Public Health, "U.S. Vital Statistics Show Death Rates Down, Birth Rates Up," January 2000, http://sdearth times.com/et0100//et0100s7.html (2/6/03).

237   ". . . Homicides . . .": U.S. Census Bureau, "Table No. 109. Death Rates by Leading Cause—States: 1997," Statistical Abstract of the United States: 2001 (121st edition), Washington, D.C., 2001.

237   "AIDS": Ibid.

237   "Car accidents": TIME Almanac 2003 (Boston: Family Education Network, 2002), p. 134.

237   ". . . Suicides . . .": U.S. Census Bureau, "Table No. 109. Death Rates by Leading Cause—States: 1997."

237   "Among cancers, lung . . .": "American Cancer Society, Estimated New Cancer Cases and Deaths by Gender, US 2002," Cancer Facts & Figures 2002, http://www.cancer.org/downloads/stt/cancerfacts& figures2002tm.pdf (1/21/03).

237   "And smoking causes . . .": American Lung Association, "State of Tobacco Control: 2002."

237   "According to the . . . Even apparently healthy . . .": Ibid.

237   "On average, smokers have $12,000 more . . .": Campaign for Tobacco-Free Kids, "Comprehensive Statewide Tobacco Prevention Programs Save Money," http://tobaccofreekids.org/research/ factsheets/pdf/0168.pdf, p. 3 (1/18/03).

237   "Federal, state, and . . .": Campaign for Tobacco-Free Kids, "Federal and State Annual Tobacco Costs and Revenues," May 17, 2000, http://tobaccofreekids.org/research/factsheets/pdf/0108.pdf.

238   "And yet, at . . ." Campaign for Tobacco-Free Kids, "Toll of Tobacco in the United States of America," http://tobaccofreekids.org/research/factsheets/pdf/0072.pdf (2/10/03).

238   "One in three . . .": Ibid.

238   "one twelfth grader in three smokes": NIH News Release, "Report Shows Recent Progress in Decreasing Youth Tobacco Use, but Much Work Remains," April 2, 2002, http://www.nih.gov/news/pr/apr2002/nci-02.htm (1/29/03); http://books.nap.edu/html/state_tobacco/, p. 1 (1/28/03).

239   "Turn off one spigot . . . Stop the ads . . .": Attorney General Michael Moore, conversation with the author, 1997.

240   "He never stood taller . . .": "Tobacco Settlement; Tobacco Chronology," USAToday.com, June, 25, 1997, http://www.usatoday.com/news/smoke/smoke26.htm, and Gene Borio, "The History of Tobacco Part IV," Historynet.com, http://www.historian.org/bysubject/tobacco4.htm.

240   "I just don't like . . .": Erskine Bowles, conversation with the author, June 1995.

240   "Smokers, even tobacco . . .": Al Gore, conversation with the author, June 1995.

240   "labeled virtual drug dealers . . .": "Jackhammer on Tobacco Road: Clinton's Civil Suit Against the Industry Has Potholes," *Christian Science Monitor,* September 30, 1999, http://csmweb2.emcweb.com/durable/1999/09/30/p10s1.htm (2/21/03).

241   "California voters had . . . ballot measure . . . As a result . . .": "California: Behind the Scenes," American Lung Association Action Network, http://lungaction.org/reports/state-narrative.tcl?geo%5farea%5fid=06 (1/22/03).

241   "In California, lung cancer rates . . .": Campaign for Tobacco-Free Kids, "Comprehensive Statewide Tobacco Prevention Programs Save Money," http://www.tobaccofreekids.org/research/factsheets/pdf/0168.pdf (1/21/03).

241   "Cigarette consumption fell 58 percent . . . More than 1.3 . . .": "Comprehensive Statewide Tobacco Prevention Programs Effectively Reduce Tobacco Use," Campaign for Tobacco-Free Kids, http://www.tobaccofreekids.org/research/factsheets/pdf/0045.pdf, p. 1 (1/21/03).

242   "Half as many . . .": Ibid., p. 2.

242   "A recent evaluation . . .": National Cancer Policy Board, Institute
       of Medicine and National Research Council, "State Programs Can
       Reduce Tobacco Use," p. 2 (1/28/03).

242   "In Minnesota . . .": Loren Stein, "How to Fight Big Tobacco and
       Win," *Consumer Health Interactive* as reprinted at *Principal
       Health   News,*   http://www.principalhealthnews.com/topic/
       bigtobacco (2/22/03).

243   "Oregon achieved similar results . . .": Campaign for Tobacco-Free
       Kids, "Comprehensive Statewide Tobacco Prevention Programs
       Effectively Reduce Tobacco Use," p. 4.

243   "Maine cut high-school student smoking by one-third": Ibid., p. 5.

243   "in Mississippi . . . it dropped by a quarter": Ibid., p. 4.

243   "CRUSH PROOF BOX": NoTobacco.org, "Cool Photo Gallery,"
       http:// www.notobacco.org/photos/index.htm (2/10/03).

243   "I started smoking . . .": Agoodmanonline, "Coming Apart at the
       Theme,"   November   2001,   http://www.agoodmanonline.com/
       newsletter/2001_11.htm.

244   "Like the political coward . . .": Campaign for Tobacco-Free Kids,
       "Big Tobacco Wins, Kids and Taxpayers Lose as California Budget
       Slashes Tobacco Prevention and Fails to Increase the Cigarette
       Tax," http://tobaccofreekids.org/Script/DisplayPressRelease.php3?
       Display=535 (2/4/03)

244   "So Davis pledged . . . then, adding insult . . .": "Budget Undercuts
       Anti-Smoking Program," American Lung Association, October 22,
       2002, http://www.californialung.org/advocacy/02fw_budget.html
       (2/4/03).

244   "Then, to top it off . . .": "Big Tobacco Wins, Kids and Taxpayers
       Lose as California Budget Slashes Tobacco Prevention and Fails to
       Increase the Cigarette Tax."

244   "In Arizona, Governor Jane Dee Hull . . .": Campaign for
       Tobacco-Free Kids, American Lung Association, American Cancer
       Society, and American Heart Association, "Show Us the Money:
       A Mid-Year Update on the States' Allocation of the Tobacco
       Settlement   Dollars,"   http://www.tobaccofreekids.org/reports/
       settlements/2002/fullreport.pdf, p. 3 (1/21/03).

244   "Fortunately, in November 2002 . . .": Arizona Department of
       Revenue, "Important Notice to All Licensed Distributors,"
       November, 14, 2002, http://www.revenue.state.az.us/tobacco/
       prop303.pdf (1/21/03).

244   "Last year, Governor Ryan . . .": Campaign for Tobacco-Free Kids, American Lung Association, American Cancer Society, and American Heart Association, "Show Us the Money: A Mid-Year Update on the States' Allocation of the Tobacco Settlement Dollars," p. 4.

245   "Under Governor Angus King . . . Indiana, Maryland, Alaska . . .": Campaign for Tobacco-Free Kids, American Lung Association, American Cancer Society, and American Heart Association, "Show Us the Money: A Mid-Year Update on the States' Allocation of the Tobacco Settlement Dollars."

245   "the American Lung Association has ranked . . .": American Lung Association Action Network, "States Ranked by Tobacco Prevention and Control Spending," Tobacco Prevention and Control Spending, January 7, 2003, http://lungaction.org/reports/rank-states.html (1/21/03).

246   "The Antismoking Honor Roll . . .": Ibid.

246   "In its fourteen years . . .": "Saving Lives—Tobacco Prevention Works," Tobaccofreekids.org, http://tobaccofreekids.org/saving lives/.

247   "Number of Lives Each State . . .": American Lung Association Action Network, "States Ranked by Tobacco Prevention and Control Spending," and Electiondb.com, "Past U.S. State Governors," http://www.electiondb.com/Historical_Govs.asp.

248   "More than half of all childbirths . . .": Campaign for Tobacco-Free Kids, "Comprehensive Statewide Tobacco Prevention Programs Save Money," p. 1.

248   "When parents don't smoke . . .": Ibid.

248   "Smoking-related fires . . .": Ibid., p. 2.

248   "For every dollar spent . . .": Ibid., p. 1.

248   "The tobacco companies increased their marketing . . .": Campaign for Tobacco-Free Kids, American Lung Association, American Cancer Society, and American Heart Association, "Show Us the Money: A Mid-Year Update on the States' Allocation of the Tobacco Settlement Dollars," p. 7.

248   "Efforts to prevent . . .": U.S. Department of Health and Human Services, Reducing Tobacco Use: A Report of the Surgeon General—Executive Summary (Atlanta, Georgia: U.S. Department of Health and Human Services, Centers for Disease Control and Prevention, National Center for Chronic Disease Prevention and Health Promotion, Office on Smoking and Health, 2000), http://www.cdc.gov/tobacco/sgr/sgr_2000/execsumm.pdf, p. 1 (1/21/03).

249 "Market share among . . .": "Federal Trade Commission Cigarette Report For 2000," Table 2—Per Capita Domestic Cigarette Sales, http://www.ftc.gov/os/1998/9803/tables96cigrpt.pdf.

249 "They're advertising . . .": Campaign for Tobacco-Free Kids, "Philip Morris and Targeting Kids," October 8, 1999, http://tobaccofreekids.org/research/factsheets/pdf/philipmorris.pdf

249 "After all, children under eighteen . . .": Mascot Coalition, "Teen Tobacco Use: Prevalence," http://www.mascotcoalition.org/education/facts/youth.html (2/22/03).

249 "tobacco companies gave the senators . . .": Green, *Selling Out,* p. 178.

249 "In all, the tobacco companies spent . . .": Ibid.

249 "Here are the names of the forty-two senators . . .": Bill # S1415, as quoted from Thomas, "Legislative Information on the Internet," http://thomas.loc.gov/ (2/5/03).

250 "In his 2000 campaign . . .": Green, *Selling Out,* pp. 178–79.

251 "The federal and state governments . . .": Campaign for Tobacco-Free Kids, "Immorality and Inaccuracy of the Death Benefit Argument," http://www.tobaccofreekids.org/research/factsheets/pdf/0036.pdf, p. 2 (1/21/03).

251 "Using the Irish . . .": Jonathan Swift, *A Modest Proposal and Other Stories* (Prometheus Books, Amherst, N.Y., 1994).

251 "Put more bluntly . . .": Campaign for Tobacco-Free Kids, "Immorality and Inaccuracy of the Death Benefit Argument," p. 1 (1/21/03).

252 "In 1999, Mayola Williams . . .": Barry Meier, "Jury Awards $81 Million to Oregon Smoker's Family," *New York Times,* March 31, 1999.

252 "In June 2001 . . .": James Sterngold, "A Jury Awards a Smoker with Lung Cancer $3 Billion from Philip Morris," *New York Times,* June 7, 2001.

252 "A Kansas jury . . .": Virginia GASP Home Page, "Reprehensible Actions—Kansas Decision"; excerpts from Bloomberg, Update 4, June 21, 2002, headlined "R. J. Reynolds Told to Pay $15 million to Kansas Smoker," William McQuillen, http://www.gasp.org/Kansas.html.

252 "In October 2002 . . .": "National Briefing—West: California: Accepting $28 Million in Tobacco Suit," *New York Times,* December 25, 2002.

252   "Research has shown . . .": National Cancer Policy Board, Institute of Medicine and National Research Council, "State Programs Can Reduce Tobacco Use," p. 6 (1/28/03).

252   "Lynn French, a fifty-six-year-old . . .": Greg Winter, "Jury Awards $5.5 Million in a Secondhand Smoke Case," *New York Times,* June 20, 2002.

253   "Even more troubling to the tobacco giants . . . In the meantime . . .": Mark Gottlieb, "Florida Jury Awards Dying Smoker $37.5 Million in Trial Stemming from Historic Engle Class Action," Tobacco on Trial, http://www.tobacco.neu.edu/tot/11-2002/engle-06-2002.html (2/23/02).

253   "Before he left office . . .": "Justice Dept. Suing Tobacco Firms," USAToday.com, September 22, 1999, http://www.usatoday.com/news/smoke/smoke287.htm.

253   "Amazingly—and to his credit . . .": David Johnston, "In Shift, U.S. Opens Effort to Settle Tobacco Lawsuit," *New York Times,* June 20, 2001.

253   "Federal and state . . .": Campaign for Tobacco-Free Kids, "Federal and State Annual Tobacco Costs and Revenues," May 17, 2000, http://tobaccofreekids.org/research/factsheets/pdf/0108.pdf.

253–   "Suppose I contributed $1,000 . . . .": Dr. David O. Lewis, Editorial, "The Hypocrisy of Big Tobacco's Advertising," *Roanoke Times*
254   Online, Roanoke.com, August 2, 2001.

255   "Health experts estimate that for each 10 percent . . .": American Lung Association, "State of Tobacco Control: 2002."

255   "In 2002, eleven . . .": Ibid.

255   "Massachusetts now has . . . $1.51 per pack . . .": Ibid.

255   "New York City smokers have the highest . . .": Patrick Fleenor, "Cigarette Taxes, Black Markets, and Crime: Lessons from New York's 50-Year Losing Battle," Cato Policy Analysis No. 468, February 6, 2003, http://www.cato.org/pubs/pas/pa-468es.html.

255   "All told, it costs $7 . . .": "Sticker Shock for Smokers?" CBSNews.com, February 20, 2002, http://www.cbsnews.com/stories/2002/02/20/national/main330058.shtml.

255   "In states that allow referenda . . .": American Lung Association, "State of Tobacco Control: 2002."

256   "Secondhand smoke contains . . . Causes pneumonia, ear infections . . .": Ibid.

256   "Forty-three states are rated F . . .": American Lung Association Network, "States Ranked by Laws Ensuring Smokefree Air," http://lungaction.org/reports/rank-states.html.

257   "California practically bans . . . Vermont bans them entirely . . .": American Lung Association Action Network, "State of Tobacco Control: 2002."

Chapter 9

259   "the plight of the 3.5 million people . . ." Christopher H. Schmitt, "The New Math of Old Age—Why the Nursing Home Industry's Cries of Poverty Don't Add Up," USNEWS.com, Health & Medicine, http://www.usnews.com/usnews/nycu/health/articles/020930/30homes.htm (9/30/02).

259   "seventeen thousand nursing homes . . .": U.S. House Committee on Government Reform Report, "Abuse of Residents Is a Major Problem in U.S. Nursing Homes," July 30, 2001, p. 4.

259   "Nine nursing homes out of ten . . .": Summary of August 2000 Health Care Financing Administration Study, Report to Congress, "Appropriateness of Minimum Staffing Ratios in Nursing Homes."

259   "In one nursing home in three . . .": U.S. House Committee on Government Reform Report, "Abuse of Residents Is a Major Problem in U.S. Nursing Homes," p. i.

259   "The abuse violations . . .": Summary of July 2001 U.S. House Committee on Government Reform Report, "Abuse of Residents Is a Major Problem in U.S. Nursing Homes."

260   "Medicaid and Medicare . . .": Robert Pear, "9 out of 10 Nursing Homes in U.S. Lack Adequate Staff, a Government Study Finds," New York Times, February 18, 2002.

260   "Residents and their . . .": Schmitt, "The New Math of Old Age."

260   "The nursing home industry . . .": Steve Vancore, e-mail to author, February 20, 2003.

260   "get paid regardless . . .": Ibid.

260   "2.2 million people die . . .": Phillip O'Connor and Andrew Schneider, "Thousands Are Being Killed in Nursing Homes Each Year," St. Louis Post-Dispatch, October 31, 2002.

261   "In Florida, for example . . .": Fox, Grove, Abbey, Adams, Byelick, & Kiernan, L.L.P., "Nursing Home Tort Reform," Nursing Home Newsletter, Summer 2000.

262   "An eighty-year-old . . . Maybe she fell on a broomstick . . .": U.S. Committee on Government Reform Report, "Abuse of Residents Is a Major Problem in U.S. Nursing Homes," pp. 10–14.

263   "Widespread incidents were reported . . .": Ibid.

263   "cited for an abuse violation . . .": Ibid., p. i.

263   "not reported promptly . . .": United States General Accounting Office—Report to Congressional Requesters, "Nursing Homes: More Can Be Done to Protect Residents from Abuse," March, 2002, p. 4.

264   "found several reasons . . .": Robert Pear, "Unreported Abuse Found at Nursing Home," *New York Times,* March 3, 2002.

264   "Patients and their relatives . . . In some states and at some nursing homes . . .": Ibid.

264   "There is no federal statute . . .": United States General Accounting Office—Report to Congressional Requesters, "Nursing Homes: More Can Be Done to Protect Residents from Abuse," p. 5.

264   "similar offenses, such as child abuse . . .": Ibid., p. 17.

265   "Our work also . . .": Ibid., p. 16.

265   "since January, 1999 . . .": Pear, "Unreported Abuse Found at Nursing Home."

265   "these deficiencies rarely . . .": United States General Accounting Office—Report to Congressional Requesters, "Nursing Homes: More Can Be Done to Protect Residents from Abuse," p. 17.

265   "Federal investigators studied . . .": Ibid., p. 20.

266   "Between 1999 and 2001 . . .": U.S. House Committee on Government Reform Report, "Abuse of Residents Is a Major Problem in U.S. Nursing Homes," p. 6.

266   "The 158 cases the GAO studied . . .": United States General Accounting Office—Report to Congressional Requesters, "Nursing Homes: More Can Be Done to Protect Residents from Abuse," p. 23.

266   "If this record of abuse . . .": Steve Vancore, e-mail correspondence with author, February 20, 2003.

266   "After performing a complex . . .": Summary of August 2000 Health Care Financing Administration Study, Report to Congress, "Appropriateness of Minimum Nurse Staffing Ratios in Nursing Homes," p. 6.

267   "experience bedsores, malnutrition . . .": Robert Pear, "9 Out of 10 Nursing Homes in U.S. Lack Adequate Staff, a Government Study Finds," *New York Times,* February 18, 2002.

267   "nursing homes with low staffing . . .": HCFA Report, "Appropriateness of Minimum Staffing Ratios in Nursing Homes," August 2000, Executive Summary.

267   "The report calculated . . .": Pear, "9 Out of 10 Nursing Homes in U.S. Lack Adequate Staff, a Government Study Finds."

267   "The Bush Administration said . . .": Ibid.

268   "The government admits . . .": quoted in Ibid.

268   "a study by the Department of Social and Behavioral Sciences . . .": Charlene Harrington, Ph.D.; Helen Carillo, M.S.; Valerie Wellin, B.A.; Baleen B. Shemirani, B.A., "Nursing Facilities, Staffing, Residents, and Facility Deficiencies, 1995 Through 2001," Department of Social and Behavioral Sciences–University of California, August 2002, p. 44.

269   "17 percent of nursing homes . . .": Ibid., p. 94.

269   "Almost one-third of all nursing homes . . .": Ibid., p. 104.

269   "It's not uncommon . . .": "Scrimping on Care," USNEWS.com, Health & Medicine, 9/30/02, http://www.usnews.com/usnews/nycu/health/articles/020930/30homes.b2.htm.

269   "Studies of nursing homes . . ." Harrington, Carillo, Wellin, and Shemirani, "Nursing Facilities, Staffing, Residents, and Facility Deficiencies, 1995 Through 2001," p. 88.

270   "In 2001, only 8 percent . . .": Ibid., p. 92.

270   "Studies show that one in ten nursing home . . .": Ibid., p. 100.

270   "But 13 percent of nursing homes . . .": Ibid., p. 96.

270   "Given these deficiencies . . .": Phillip O'Connor and Andrew Schneider, "Thousands Are Being Killed in Nursing Homes Each Year," *St. Louis Post-Dispatch,* October 13, 2002.

271   "I didn't know . . . These poor souls . . .": quoted in Ibid.

271   "With government and private sources . . .": Schmitt, "The New Math of Old Age."

272   "Ranking of States . . .": Harrington, Carillo, Wellin, and Shemirani, "Nursing Facilities, Staffing, Residents, and Facility Deficiencies, 1995 Through 2001," p. 61.

273   "exhausting, unpleasant, and often dangerous . . .": "Scrimping on Care," USNEWS.com.

273   "According to the . . .": Phillip O'Connor and Andrew Schneider, "Woefully Inadequate Staffing Is at the Root of Patient Neglect," *St. Louis Post-Dispatch,* October 14, 2002.

273   "Two-thirds of America's nursing homes . . .": Harrington, et al. "Does Investor Ownership of Nursing Homes Compromise the Quality of Care?" *American Journal of Public Health*, September 2001, Summary.

273   "Nursing homes that . . .": Ibid.

273   "privately owned, for-profit . . .": Harrington, et al., "Does Investor Ownership of Nursing Homes Compromise the Quality of Care?" *American Journal of Public Health,* September 2001, p. 1453.

273   "Our results suggest . . .": Ibid., p. 1454.

274   "America's top nursing home chain . . .": Ibid.

274   "*U.S. News & World Report* noted . . .": Schmitt, "The New Math of Old Age."

274   "At another nursing home . . .": Ibid.

274   "improperly made to them": Ibid.

274   "self-dealing is widespread . . .": Ibid.

274   "a number of highflying . . .": "Nursing Home Firm Settles Fraud Case," *Washington Post,* February 4, 2000.

274   "whose pay was only $12–$25 . . .": Julia Malone, "Nursing Home Bankruptcies Prompt Calls for Aid, Questions About Collapse," Cox Washington Bureau, September 3, 2000.

274   "It's the smell that lingers . . .": Christopher H. Schmitt, "Home $weet Home—Big Fees," USNEWS.com, Health & Medicine, September 30, 2002.

275   "*April 1999:* Vencor Inc.": "Vencor Loss $651 Million, Feds Pursue Overpayment" Associated Press/*Lexington Herald-Leader,* April 17, 1999.

275   "*May 1999:* Integrated Health Service . . .": Chart of highest-paid CEOs vs. underpaid CEOs, *Forbes,* May 17, 1999:

275   "*February 2000:* Beverly Enterprises . . .": "Nursing Home Firm Settles Fraud Case," *Washington Post,* February 4, 2000.

275 "*January 2001:* Integrated Health . . .": "Health Chain's Growth Was Bold," *Baltimore Sun,* January 13, 2001. ($44 million salary figure from *Forbes,* cited above.)

275 "*March 2001:* Vencor Inc . . . .": "Vencor and Ventas Paying U.S. $219 Million to Resolve Health Care Claims as Part of Vencor's Bankruptcy Reorganization," U.S. Department of Justice Press Release, March 19, 2001.

275 "*October 2001:* Sun Healthcare . . .": "Nursing Home Firm Assents to Conviction," *San Francisco Chronicle,* October 5, 2001.

276 "*December, 2001:* National Healthcare . . ." "Tennessee-based National Healthcare Corporation Settles Medicare Fraud Case for $27 Million," U.S. Department of Justice Press Release, December 15, 2001.

276 "*August, 2002:* Beverly Enterprises . . .": "Nursing Home Giant Settles with State," *Santa Rosa Press Democrat,* August 14, 2002.

276 "Medicare [profit] margins . . .": Medicare Payment Advisory Committee, Public Meeting, December 12, 2002.

276 "hundreds of thousands of pages . . .": Schmitt, "The New Math of Old Age."

276 "the nursing home industry . . .": Ibid.

276 "Included was a reform . . .": Ann Imse, "Bankruptcy Hits Nursing Home—State's Frailest Citizens Put Trust in 58 Facilities Run by Ailing Companies," *Rocky Mountain News,* October 22, 2000.

277 "The cut reduced Medicare payments . . .": Ibid.

277 "One such was Integrated Health . . .": M. William Salganik, "Health Chain's Growth Was Bold," SunSpot.net Business, January 13, 2001.

277 "became known not only . . .": Ibid.

277 "Profitable in the second quarter . . .": Ibid.

277 "leveraged up the gazoo": quoted in Ibid.

277 "Five of the nation's seven . . .": Ibid.

277 "a coast to coast . . .": Schmitt, "The New Math of Old Age."

277 "In studying industry finances . . .": Ibid.

278 "The companies that have gone bankrupt . . .": quoted in Malone, "Nursing Home Bankruptcies Prompt Calls for Aid, Questions About Collapse."

279   "If the available financial estimates are correct . . .": Robert Pear,
       "9 Out of 10 Nursing Homes in U.S. Lack Adequate Staff, a Gov-
       ernment Study Finds," *New York Times,* February 18, 2002.

280   "Colorado Health Care Association director . . .": Ann Imse,
       "Bankruptcy Hits Nursing Home—State's Frailest Citizens Put
       Trust in 58 Facilities Run by Ailing Companies," *Rocky Mountain
       News,* October 22, 2000.